# USING GOOGLE'S FREE S#!T

## GROWING YOUR BUSINESS TO SIX FIGURES & BEYOND

ROBIN C. STERN

ILuvMyGeek Publications
-New York-

**Using Google's Free S#!T**
Growing Your Business to Six Figures and Beyond

Copyright© 2023, 2025 by Robin C. Stern. All rights reserved.

Published by:
ILuvMyGeek Publications
Brooklyn, NY
917-947-9633

All Rights Reserved. No part of this book may be used or reproduced in any manner whatsoever without the expressed written permission of the author. Address all inquiries to:

Robin C. Stern
Telephone: 678-685-9759
Email: robin@robincstern.com
www.UsingGooglesFreeStuff.com
www.RobinCStern.com

ISBN: 979-8-218-23831-5

Library of Congress Control Number: 2023913463

Editor: ILuvMyGeek Publications
Cover Designer: ThumbSwipe Graphix
Interior Book Layout: ILuvMyGeek Publications
Author Photo: Kevin Urbina, Media Titan

Every attempt has been made to properly source all quotes.

Printed in the United States of America

Second Edition

4 6 8 10 12 14

# WHAT OTHERS ARE SAYING ABOUT ROBIN C. STERN AND THIS BOOK

"Robin is a tech marvel with so much knowledge in so many areas. As an attendee on a couple of my early coaching calls, he spoke up during my technical frustration and had the quick answer for the issue at hand. Now after getting to know him and reading his book, I can truly say Robin C. Stern is known as YOUR Google Guru and that's because he's got a knack for making complex Google stuff totally understandable. 'Using Google's Free S#!t' is exactly what it says: a no-BS guide to using free tools to grow your business. If you're overwhelmed by online marketing, grab this book. Robin makes it simple."'"

— Forbes Riley: "Queen of Pitch", Celebrity TV and 3x Best Selling Author

"Robin C. Stern is an amazing speaker! He has shared his expertise with small businesses in our community multiple times. I highly recommend him!"

— Sandy Gamble, VP Operations, Paulding Chamber

"Robin C Stern is an awesome Google Guru. I love how he teaches with such ease. Thank you so much Robin!"

— Aida Sanchez, Founder Arise Dream Travel

“I had the pleasure of working with Robin to better understand and navigate Google which has truly made all the difference in my business. Thank you so much for your insights and help, Robin!”

— Lenora Edwards, DTB Horizons

“An expert’s knowledge for the everyman! Robin presents an abundance of material in an entertaining way. His explanations on how to utilize Google's content are incomparable and his experience is evident!”

— Benjamin, 5-star review on Amazon

“Overall, "Using Google's Free S#!t!" is a valuable resource for any business owner seeking to maximize their online presence and achieve sustainable growth. Highly recommend!”

— Joe Minor, Joe Minor Coaching

“Robin is Brilliant! I am easily overwhelmed by all the little things a modern business owner needs to do to maximize their Google results. Robin explained it all in an absorbable way and made it fun and easy for me to get the most from Google including getting me across the finish line on verification with ease. 10000 stars!”

— Kira Coopersmith, Founder, Sensible Sort

"You are not Google's customer; you are Google's product! Robin presents this exciting, interesting perspective in a fun, easy-to-read book with a checklist to guide you. Change your thinking to become Google's product to transform your business and your life!"

— Jim Hickey, Colonel US Army (Retired)

"Robin presented to us at the Paulding Chamber of Commerce. He taught me that Google is the baker, and I am the cookie. I need to be the freshest cookie in town to help gain business. His advice changed my outlook on how much time I spend on my Google profile. Thanks Robin!"

— Garrett Potz, Affinity Lending

"Robin's book breaks down free tools into simple, actionable steps that anyone can follow. His clear explanations and real-life examples make it easy to understand and use these tools effectively. This book is packed with valuable insights and practical tips."

— Charles Higgins, Founder, Pinnacle AI

"Robin is always so helpful and is a networking master! His book is an amazing tool for any business professional. He is always willing to help and gives the best advice."

— Staci Hamilton, Owner, Staci Hamilton Realty

"For most of us, getting "clicks," developing leads, and converting prospects is either a mystery or a very expensive proposition. But with USING GOOGLE'S FREE S#!T, you'll not only learn how to create and leverage your Google Business Profile, but how to get branded Email, get reviews automatically, and a whole host of other S#!T (stuff) – all free! You'll keep more of your marketing dollars and get dramatically better results that will allow you to take your business to the next level – Your Business 2.0."

— Larry Zgrabik, GPS Consulting, Robin's First Biz Coach

"Very helpful with his suggestions on what I can do to get my business unsuspended from Google. He gave me lots of suggestions on what to try. Really appreciated his help."

— Rachael Venema, Rachael Venema Photography

"Robin's book is exactly like Robin - packed full of value that we only receive because we connect with Robin C. Stern. Get the book and allow it to make you more "Google Effective!" And if you ever get the chance to sit with Robin over coffee, take it! You will be glad you did!"

— Keith Ivey, Keith Ivey Hypnotherapy

"Robin was so helpful! My problem was fixed in 5 minutes flat. Thank you SO MUCH!!!!"

— Julie Collins, Judy Collins Photography

"His book was so insightful! It helped me navigate Google and get a lot of free things that I didn't even know Google offered. I have recommended this book to almost every business owner that I know. You won't be disappointed with purchasing this book it has true value in it."

— D'Naya Eubanks, Founder, Beneath, Inc.

"I just attended a presentation by Robin C Stern, Your Google Guru, and what I learned is amazing! I can't wait to use what I learned! Make sure to buy the book also! You won't be disappointed!"

— Michelle Lennard, Aflac

"Want to increase your business by 10X? Start here!"

— Cindy Lee Rhodes, Author and Bucket List Coach

"This book is the best investment you can make in your business!"

— Patrick Snow, Publishing Coach and Best-Selling Author of *Creating Your Own Destiny* and *The Affluent Entrepreneur*

"I don't care how much money you have;
free stuff is always a good thing."

— Queen Latifah

# DEDICATION

To my wife Elaine Stern:

You are the love of my life. Thank you for always believing in me and putting up with me. You came into my life at a time when I thought every good thing was gone. You showed me that wasn't the case. Thank you.

To my sons Tyler Stern, Cristian Stern and Trevor Stern:

I could not imagine life without you. I pray you all have a deep sense of who you are and how much you are loved.

To my dad Joe Stern (1928 – 2010):

I have always known I could do anything and that is because you showed me how. You have always been my superhero.

To my mom Anne Stern (1930 – 1993):

You told me I was the most amazing person in the world, and because of you I am.

To you my reader:

Growing your business to six figures and beyond is available to you. I am hopeful that the information in this book will help you to achieve your goals and dreams for your business, your family, and your future. Godspeed!

"I am always doing what I cannot do yet, in order to learn how to do it."

– Vincent van Gogh

# ACKNOWLEDGMENTS

To all my friends, family, and business associates, especially the ones who purchased this book before it was even finished, thanks for believing in me.

Forbes Riley, Google, Alphabet Inc., Gwen Epley, ESPN, Catherine Herd, Angela Sims, Cheryl Hyde, Robert Becker, Alyssa Schwab, Joel Saxe, David Feldstein, Shatara Hawkins, James Lloyd Hickey, James P. Lubbe, Darlene Van Beek, Tiffany Miller, Jerry Bohus, Julie Groom, Kimberly George, Amy Novick, Gary Vaynerchuk, Cindy Lee Rhodes, Me, Ronald VanGelder, Giovanni Gonzalez, Larry Zgrabik, Dad, Ryan Reynolds, Stacy Hamby, Patricia Stern, Jan Britt, Joanna Satterwhite, Saurel Quettan, CeCe Carr, Tara Heaton, Dr. Richard Block, Kathy Lathem, Susan Stern, Tyler Stern, Fred Roberto, Sarah Salazar, Kristen Russo, Kim Roberto, Scott Satterwhite, Yolanda Newell, Charles Higgins, J Lo, Mary Snider, Sandy Gamble, Shelbia Bone, Pixar Studios, Elaine Stern, Janet Stern, Yeshua, Luciano and Nina Mena, Greg Sanders, Vince Baker, Mike Hayes, Anastasia Kudrez, Andrea Villa, Andrew and Angelica Rodriquez, Tyrik Wynn, Jim T. Chong, Blake Jones, Bob Brooks, Carl Bromley, Charlie Darrien, Chris Lambert – USMC, Cliff McGrady, Boyd Monk, Warren Buffet, Chris Tice, Angelia Farries, Danny Camp, Dave Roberts, Oprah Winfrey, Elon Musk, Foo Torres, Rupert Murdoch, Jeff Bezos, Sundar Pichai, Charles LeCroy, Lori Kingery, Alissa Pounds, Katy Grainger, Mary Rose Campbell, John Wise, Dena Humphries, and to anyone I missed, my apologies. You know who you are.

# SPECIAL ACKNOWLEDGMENTS

To Google and Alphabet Inc., the brilliant minds and innovative spirit behind the platform that has become the very foundation of my work and, dare I say, my persona as "Your Google Guru."

Without the groundbreaking tools and resources you've so generously made available – let's be honest, it's practically FREE magic – this book, "Using Google's Free S#!t" would simply not exist. Thank you, from the bottom of my heart (and my ever-overflowing Google Drive), for empowering me and countless others to learn, create and connect in ways previously unimaginable. You've not just built a platform, you've built a launchpad for dreams.

This book stands as a testament to the transformative power of your innovations, which have not only democratized access to information and technology but has also enabled individuals like me to flourish as educators and guides in this ever-evolving digital landscape. For your unparalleled contribution to the world and your indirect yet essential partnership in this literary endeavor, I offer my sincerest thanks and deepest admiration.

G
OOG
LEBU
SINESS
PROFILE

# CONTENTS

"The journey of a thousand miles begins with a single step."

—Lao Tzu

# PREFACE:
## BEING THE FRESHEST COOKIE

"As you all know, first prize is a Cadillac El Dorado. Anyone wanna see second prize? Second prize is a set of steak knives. Third prize? Third prize is you're fired!"

—From the Movie, Glengarry Glen Ross

How would you describe your relationship between Google and your business? Do you use it strategically? Sorta just hope for the best?

This is where most business owners get it wrong. When managing your Google Business Profile:

**You are not Google's customer—you are Google's product.**

Let that sink in. See how that insight shifts the way you should approach your presence on Google? How it redefines the way you engage with your Google Business Profile?

Let's break it down with an example.

A customer walks into a bakery, clearly ready to buy. The baker sees the opportunity; there are plenty of delicious options within reach, and he knows each one inside and out. His goal? To impress the customer, make him happy, and turn him into a loyal fan who not only returns but also tells all his friends.

"I am looking for some great cookies" says the customer.

The baker knows he has fresh cookies in the display case—each one a great representation of his bakery's quality. But he wants to go beyond good; he wants to wow this customer. So, he steps into the back and returns with a warm, freshly baked chocolate chip cookie, straight from the oven. The customer is impressed, becomes a loyal regular, and spreads the word to his friends, who start coming in too. The baker is happy!

In this scenario, which part do you think represents your business?

1. The customer?
2. The baker?
3. The other bakeries in town?

Well, you are not the baker—Google is. Google has someone (the customer) searching for something (the cookies) and knows exactly where to find it. But the baker decides which cookie the customer gets to sample first. So where does that leave you?

Your Google Business Profile is the cookie. Yeah, I said it. *You are the cookie.* And not just any cookie. You need to be the *freshest* cookie. Because in Google's bakery, the freshest cookie gets placed front and center. The freshest cookie wins first place in search.

That's the secret sauce. You have to signal to Google that your business is warm, fresh, and packed with the right ingredients so that it chooses *you* as the standout option. So that you're the business that gets noticed and the business that keeps customers coming back for more.

Ok, you can't put yourself on a cookie sheet and pop yourself in the oven at 375°F for 10 to 12 minutes so how do you become the freshest cookie in the bakery?

This is part of the Guru's secret. Becoming and staying the freshest cookie in your whole area is what *YOUR Google Guru* will teach you in the following chapters.

And you might just learn about some really cool Free S#!t that can save your business thousands of dollars every year.

Following each chapter I will be asking you to reflect on what you just read and how it relates to your business. This is meant to get you to really think about how you are promoting your business and how you can make some changes to better you daily operations and marketing.

1. How does learning you are Google's product and not their customer make you feel?

______________________________________________

______________________________________________

______________________________________________

______________________________________________

______________________________________________

2. If you could do one thing today to get your business seen more, what would you do?

______________________________________________

______________________________________________

______________________________________________

______________________________________________

______________________________________________

3. When was the last time you worked on or posted an update to your Google Business Profile?

______________________________________________

______________________________________________

5. Are you ready to start getting your business found more online? Do you want to leverage potentially thousands of dollars of advertising for no more cost other than your time investment?

________________________________________

________________________________________

4. What is your favorite bakery item?

________________________________________

________________________________________

Dang, now I want cookies…...

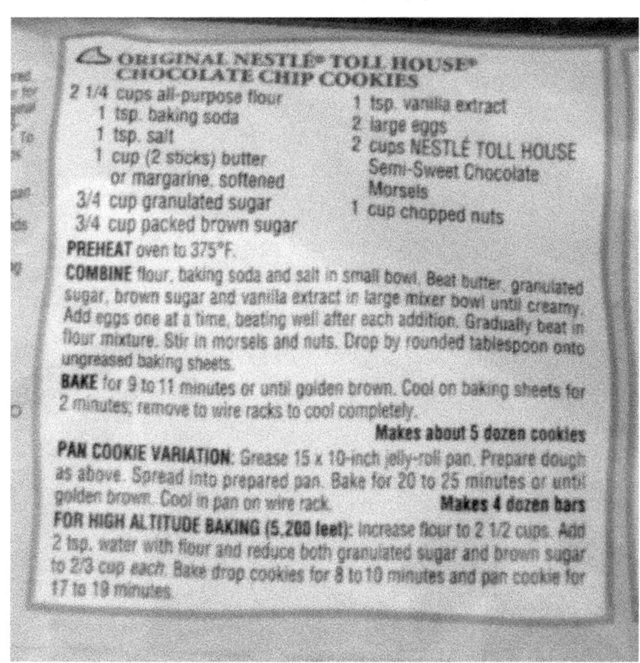

This edition of our book has a digital side. At the end of each chapter there will be a QR code to link you to our video page to watch the latest updated information for that chapter. This will allow us to stay updated on what Google is doing without having to print new books each time they change something. One more added value from *YOUR* Google Guru!!

Look for the QR code!

"Do the best you can until you know better, when you know better, do better."

— Maya Angelou

# INTRODUCTION:
## GETTING ON THE GOOGLE BANDWAGON

You are out of your mind if you're not using Google for your business! Without it, you're throwing away perfectly good money without even realizing it. You're paying for social media ads and direct mail that aren't making the phone ring or bringing in customers. Your website isn't getting the traffic you were promised. And to top it off, the calls keep coming—telling you your Google listing is wrong, even though you just paid someone to set it up!

It's no wonder you're frustrated. Every day, you're bombarded with internet marketing noise—everyone telling you to spend more money, use the latest strategy, or trust another so-called expert. SEO, analytics, algorithms—just hearing these buzzwords might make you want to scream. And yet, every time you ask for advice, the answer is the same: *Just Google it.*

Don't get mad at me...but the best place to promote your business every single day is...well, Google. Think about it—when you need to buy a product or hire someone, where do you look first? Wasting time and resources on scattered online marketing efforts is exhausting, but let's get real: if you're not prioritizing Google, you're a fool! 92% of the world turns to Google when they're ready to buy. The only question is, will you show up?

I feel your pain. I have been in your shoes. In fact, I wore out a couple of pairs of shoes promoting my first business (before Google), so I really get it. I understand your situation in that you have been so busy with life, parenting, relationships, paying bills and running a business that you simply have not had the time or the bandwidth to fully leverage Google's Free S#!t to apply to your daily business operations. And that is okay because now you have come to the right place.

In this book you will learn:

What is a Google Business Profile?
Why do you need a Google Business Profile?
What can a Google Business Profile do for your company?
How do you set up your Google Business Profile?
How does Google view your company?
How can you get more algorithm love?
What is SEO all about?
When should you use Google Ads?
What specific actions will help you stay on top of the Google Search results?
What actionable items will *help YOU stay motivated!!*

If you grasp the principles in this book and apply the strategies to your business, you'll be well on your way to leveraging Google's free S#!t to reach six figures and beyond. By making the most of the tools you're likely underutilizing, you could add an extra $1,000 to $2,000 a month to your business account—without spending a dime.

I have been a power user of Google products for more than 16 years. Leveraging every free product and service from Google has helped my companies save over one million dollars combined!

With more than 40 years of marketing and advertising experience, I have a keen eye for what works and what gets customers' attention. I started using Google Business to promote my computer company, iluvmygeek.com and I have leveraged that Business Profile into more than 580 reviews and a 5-star rating for a brick-and-mortar computer store. The local computer repair market in my area used to have 6 computer repair businesses. There are two now. (a local Best Buy electronics store just closed and Simply Mac has filed for bankruptcy, but iluvmygeek.com claims nothing to do with that...)

I have helped thousands of businesses increase their organic page ranking on Google and many of those have gotten on the first page of the search results without using paid ads. I have presented at local business events as well as internationally at various summits across the globe, helping all sized businesses to increase their visibility and reduce their operational costs by leveraging all the available Free S#!t from Google. This has led many of my clients to tag me as their Google Guru, and the name has just stuck. Now I want to be *YOUR* Google Guru and help your business grow!

My professional success has allowed me to focus my energy on giving back by offering weekly computer and internet classes for seniors at the local library, as well as serving as the chairman of Paulding County's College and Career Academy Cybersecurity Pathway Advisory Board.

I started the non-profit Be Better Not Bitter in 2016 after the death of my closest friend Ron VanGelder. This organization collects donated computers, repairs them, and then distributes them to deserving students and families around the world. Our technology training program March4Tech teaches teenagers how to repair computers, giving them first-hand experience in a much-needed career field.

I have now directed my passion for helping others toward supporting you and your company. The internet marketing, website, and advertising fields change daily. There is no way any business owner can keep up on their own; however, if you take the time to learn why you need these things and then have a trusted source to get the information, you can and will be more successful. The world needs your great ideas, so what are you waiting for!?
Will your business show up?

I can help your company get found more online and guide you through the ins and outs of Google Business and the world of online marketing. I'll break things down clearly, leaving you with the knowledge to confidently navigate the chaos of internet marketing. Plus, I'll show you how to keep some extra $$ in your account so you can celebrate your success in style.

Ready to get started? Ready to think outside the box? Ready to see how simple this can really be? Great. Then let's go, and let's begin this journey together.

I wish you good luck in the process!

1. How much money do you spend on advertising? Really think about this question. Signs. Vehicle lettering. Boosting posts on social media. Your website cost. Branded email. Direct mailings. Flyers and printed matter to give away at events. Make notes about some of the costs that come to mind and refer back as you think of more.

_______________________________________________

_______________________________________________

_______________________________________________

_______________________________________________

_______________________________________________

2. Besides buying this book, when was the last time you spent money on something to help you promote your business?

_______________________________________________

_______________________________________________

_______________________________________________

_______________________________________________

_______________________________________________

3. Knowing that 92 % of the world is using Google to search for the stuff they need, what do you think your Google Business Profile could be worth to your company if you showed up at the top of the search results every time?

_______________________________________________

_______________________________________________

_______________________________________________

_______________________________________________

4. Are you uncomfortable with technology and worried you might mess something up?

______________________________________________

______________________________________________

______________________________________________

______________________________________________

______________________________________________

5. Do you spend any amount of time or money promoting your business in your local market? List any local events or meetings you go to.

______________________________________________

______________________________________________

______________________________________________

______________________________________________

______________________________________________

______________________________________________

______________________________________________

______________________________________________

______________________________________________

______________________________________________

______________________________________________

______________________________________________

______________________________________________

6. Would you like more free time to spend with your family and to have a larger impact serving your local community? What would that look like? List 3 things you would do if you had more money and free time.

Remember, for updated information about this chapter, scan the QR code and watch the video for this chapter!!

One of my first Google Business clients was as non-technical as they come. He wasn't interested in all this "newfangled stuff!" But, based on his situation and my encouragement, he was willing to give it a try. Watching this underdog rise from obscurity to dominate the local market has been incredibly rewarding. Even more inspiring is seeing his company fund future missions and charitable work. I'll share his story throughout this book to motivate and encourage you—if George can do it, *anyone* can!

George's Story:

> "I got 2 ex-wives, 3 child support payments and at least that many lawyers who all want a piece of me Rockstar," George declared as he sat down at my desk. "I gotta make some money quick-like."
>
> This is the beginning of George's journey to his ultimate understanding of Google Business Profile, (although it was called Google My Business back then).
>
> "Well George, I am not sure if I can help with the "quick like," but I think we can get you pointed in the right direction. Whacha thinking about?"

(continued…)

# SECTION I
# GETTING STARTED

Preparing Yourself

# LEARNING THE GROUND RULES

> "Rule No. 1: Never Lose Money. Rule No. 2: Never Forget Rule No. 1."
>
> —Warren Buffet

Before we dive in, let's set some standards. With so many ways to access the internet and so many moving parts to Google, establishing guidelines will make this book easier to follow:

## GOOGLE ACCOUNT:

You need an email address that is connected to Google so you can access all the Free S#!t from Google. If you are not familiar with what a Google Account is, you can head to Chapter 3 and learn all about Google accounts. (Hint: Gmail is a Google Account)

You need to be logged into the Google account you started your Google Business Profile with, or the one you want to start your Google Business Profile with. You can see what account you are logged into by the icon or image in the top right corner of your browser.

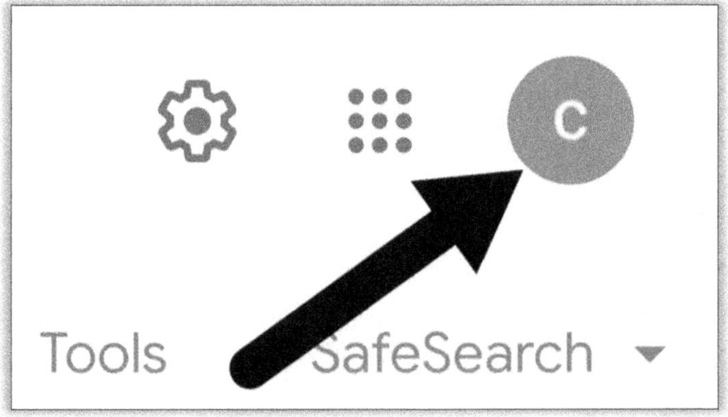

## IF YOU ALREADY HAVE A GOOGLE BUSINESS PROFILE:

Once you are sure you are logged into the right account, you will Google the words "my business" and the results page will look like a normal search results page. Upon further inspection, (and possibly a little scrolling) you will see a section that says, "You manage this Business Profile" and you should see your business listed there.

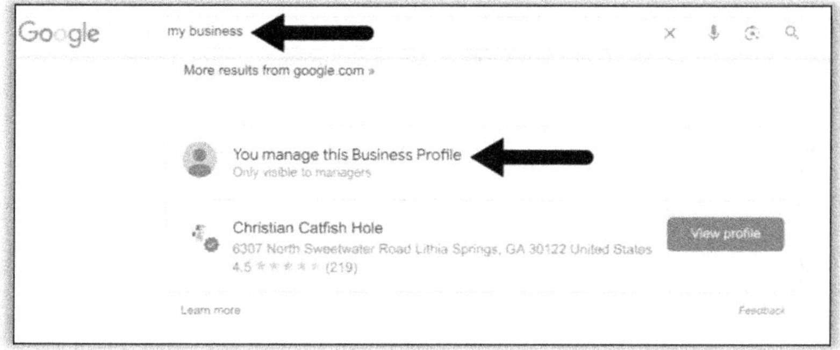


If not, there are a bunch of reasons why you might not see your business listed.

1. Did you stay logged into the correct Google account? Sometimes after you do a Google search it may change the account you are currently logged into. Check the top right to make sure you are still logged into the right account.

2. Are you sure you are using the right email address? Make sure you are using the email address that you used when you created the Google Business Profile.

3. Maybe you have an additional email address that could be connected to the business?

IF

YOU DON'T HAVE A GOOGLE BUSINESS PROFILE

OR

YOU ARE NOT SURE IF YOU HAVE A GOOGLE BUSINESS PROFILE

OR

IF YOU DIDN'T FIND YOUR BUSINESS IN THE PREVIOUS STEP:

Head to https://maps.google.com. In the top left corner of the screen, you will see a stack of 3 flat lines.

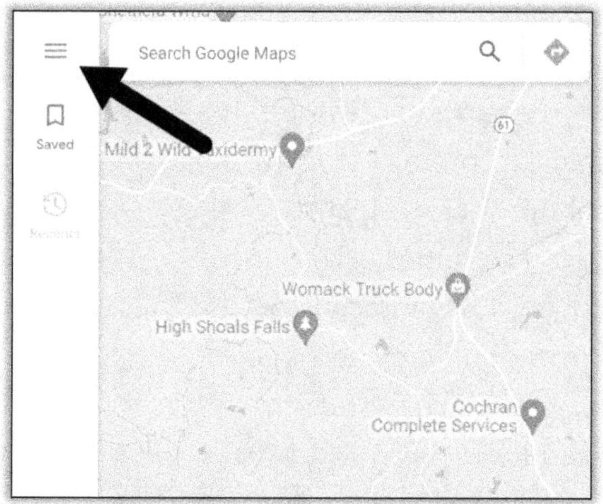


We call this a hamburger stack, and you will see that referenced in this book. If you click the hamburger stack, on the menu that appears, you will see the option to….

Add your business. Bingo! Click on that sucker and you will be directed to start the process to add your business to Google.

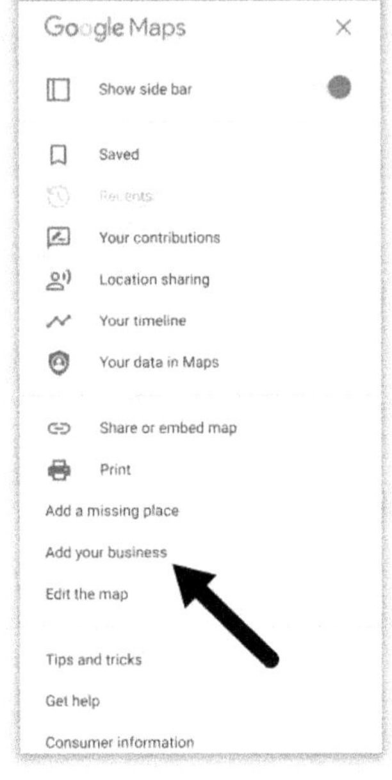


This process is straight forward. Google is going to prompt you for information about your company. At this stage you can input all your company info, but I recommend you only put in the minimum and get through the preliminary stuff as quickly as possible.
You will have plenty of time to then go through the rest of this book chapter by chapter as we discuss each section of your Google Business Profile in detail.

If you are adding your business for the first time, here are a couple of tips to help you complete the process more quickly:

- Don't worry about inputting any information that has a "Skip" button. Just get through the setup process first, and then you can go back and follow my step-by-step process in Chapter 4 to set up your Business Profile.

- Don't worry about putting photos other than your logo if you have it handy. There is a whole chapter on photos, and that covers the logo and cover photo as well as others.

- When inputting your address if you work from home, don't show your home address. Include a service area so Google will know not to send customers to your physical location. Google's official comment about home-based businesses is that they must have an office or store that clients or customers come to, AND they must have permanent signage. Unless you meet these requirements, turn off the "Show business address to customers" switch and use a service area instead.

- When you are putting in your service area, input the counties of the area you serve. You can put the area in city by city, or by zip code, but if you list the county, it will include all the cities and zip codes in that county.

If you are a home-based business, just make sure you represent your business correctly and Google will provide the customer with accurate information.

If you don't have permanent signs at your house, don't list your home address. Stick to the service area.

Once you get through the setup process, your business must be verified. Google has many ways they have been verifying businesses lately. The most common has been getting a text to the business' listed phone number (the one you just told Google was your business' number) you can have them text or call to give you a 6-digit code. Get through the verification, and your setup will be complete!

Now we must give Google some time to get a person to review it and within 24-48 hours your business will be live on Google. (Sometimes this happens faster, but the official word is 24-48 hours)

Now that you have a profile, you can jump into week one and Get Your Google On!

1. What is your default web browser?

________________________________________

________________________________________

________________________________________

________________________________________

2. What email address are you using for your Google Business Profile login? Write it here. This is your login for your Google Business Profile.

________________________________________

________________________________________

________________________________________

________________________________________

3. Google the words "My Business". Does your Google Business Profile show in the search results?

______________________________________________

______________________________________________

4. Is your Google Business Profile verified and visible publicly?

______________________________________________

______________________________________________

5. If you are a home-based business, did you tell Google not to list your address publicly?

______________________________________________

______________________________________________

______________________________________________

Remember, for updated information about this chapter, scan the QR code and watch the video for this chapter!!

Scan here to watch this chapters updated video

George's story continued:

"I wanna start a business. I got this great idea. I have a pickup truck, right? I also have a storage unit and a flea market booth.

If I can get people to call me to pick up their "junk," I can get paid for picking it up, and with the nicer stuff, I have a shot at getting paid for selling it at the flea market! Genius, right?!"

## CHECKING YOURSELF

> "… So come on and chickity-check yo' self before you wreck yo' self."
>
> —O'Shea Jackson, Sr. (Ice Cube)

All right. So, this whole "starting a business" thing isn't as easy as it seemed when writing it down on a napkin that night at the restaurant after three drinks. Maybe you have a business, and you are finally deciding to elevate your game and take that business to six figures and beyond. Either way, it is going to require you to commit to some work. However, if you believe in your idea, and you agree you are willing to be all in, then this book might help you to get your word out quickly and at a lower cost than traditional marketing methods. I'm not saying my way is the only way; I just think it is the best and least expensive.

There are a few things you must know before you jump right in to start (or grow) a business. Here is the short list:

You MUST be willing to eat ramen for months (ok, maybe not ramen, but just be aware that you will be spending less on eating and rarely eating out).

You MUST be willing to get 5 or less hours of sleep per night for an undetermined amount of time (naps and caffeine will be your friend)

You MUST trust your product or service. Do you believe in yourself? In your ability to provide a great service? Is your product the absolute best product you can offer at the best price? Are you willing to place everything you have on the line for this endeavor to become not just a reality but a raging success? Are you? If you don't believe in your idea enough to risk it all, why should anyone else?

Your business is more than just a great idea! This is how you are making your mark on the world. The bigger you grow, the bigger the mark. To grow you will need capital, and investors want to see skin in the game. This will require you to pitch your whole process, and being able to evaluate the worth of your company is a requirement for all business owners. This will include the amount of your time and personal financial resources that you are willing to invest.

You MUST be all in.

You MUST be willing to do whatever it takes.

- Read a book. (And do what it says!!)
- Don't have that drink with dinner. (You need to work a couple more hours after dinner.)
- Forgo personal needs. (Vacation, sleep, social interaction) for a period to push through to your goals.
- Get out of your comfort zone! Sell, pitch, network, cold call.
- Be creative.
- Invest everything.

There is *no reason* you cannot achieve success with plenty of sleep, good expensive dining habits and a full social calendar. It will just take you longer to get there!

You MUST have capital. It takes money to make money—the more the better. But heaps of money won't make a great business. You need to be all in financially of course, but you also must be ready to risk it all and have the balls that are necessary to make your dream come true. Where can you get capital? There are a bunch of areas. Here are a few:

- Savings
- Investors
- Personal sources
- Family
- Equity in your home
- Credit cards (short term only)

You MUST have a great name! You may think the name you have picked fits perfectly, and it very well may. It is very common to come up with the name first and then try to get all the marketing and branding to fit the name. However, be open to the idea that a different name may be more indicative of what you do and may result in better sales and search rankings.

Small changes open big possibilities for higher Google positioning.

If you already have a business name (LLC, or Corp) that doesn't reflect what you do or what you fix, there are some things you can do to stand out more. If you want to contact me for a brainstorming session to discuss that, I would love to offer any insight I may have!

1. How accurately does your business name reflect what you do?

______________________________________________

______________________________________________

2. Are you willing to do whatever it takes to make this business a success?

______________________________________________

______________________________________________

______________________________________________

3. I know habits are hard to break and changing them can be tough. You must start. What night (or morning) of the week are you willing to give up 2-3 hours of sleep to work on your business?

______________________________________________

______________________________________________

______________________________________________

4. Do you have any sources of financial support to help you get your business started? List 3 sources of capital ($$) you can access now.

______________________________________________

______________________________________________

______________________________________________

______________________________________________

______________________________________________

5. Thinking about funding/capital, do you have the resources to finance this business and survive? Have you mapped out what you will need to get past the startup stage and into profitability? Take a minute to write down some of the costs you expect to incur over the next 30 days. 60 days. 90 days.

____________________________________________

____________________________________________

____________________________________________

____________________________________________

____________________________________________

6. Are you ready to:

- Commit?
- Pay for your website and email address?
- Look like a real business?

Write down 3 things you can do today to increase the professional look and feel of your business.

____________________________________________

____________________________________________

____________________________________________

____________________________________________

____________________________________________

____________________________________________

____________________________________________

Remember, for updated information about this chapter, scan the QR code and watch the video for this chapter!!

George's story continued

George and I are talking:

"Well, that IS a great business idea George, but how do you get people to call you?"

"That's where you come in Rockstar," George replied, "you are going to get them all to call me!! You are gonna make me some signs for my truck, and maybe a website. I know you can do it Rockstar, and if you do it, it will work!"

# YOUR FIRST FREE THING: GMAIL

"Brown paper packages tied up with strings. These are a few of my favorite things…"

—Julie Andrews, from the movie
*The Sound of Music*

Without any doubt one of my favorite free things from Google is Gmail. To take advantage of all the Free S#!t from Google, you need to connect to Google. To connect to Google, you need a Google Account.

Besides Google Search, Gmail is most people's first real introduction to Google. Gmail is already by default a Google account. The G in Gmail stands for Google. Google mail. Gmail. So, if you have a Gmail account already, great!

If you don't have a Gmail address, let's look at how you can get a free Gmail address, or make your existing email address a Google account. Go to https://gmail.com and you will see this sign in screen:

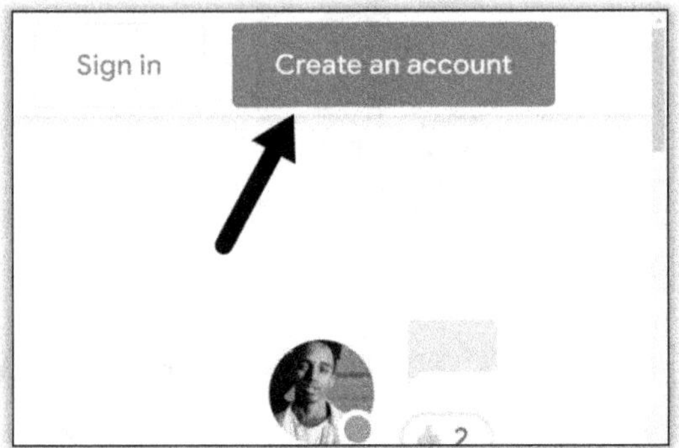


Click "Create an account" in the upper right. Then you will see this screen:

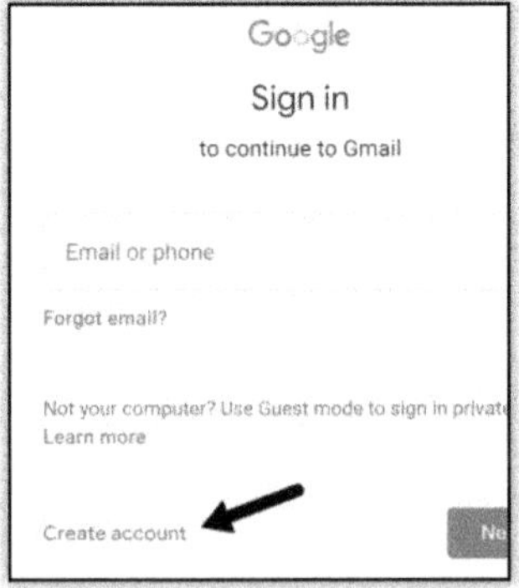


Once on this screen you will click the "Create account" link and select the option: For my personal use. This is how you will create a free Gmail address to use to access all the Free S#!t.

If you want to use your existing non-Gmail email address, you can register it with Google and sign in with it. You will need to do a few more steps:

- Google the words "use my email as a google account."

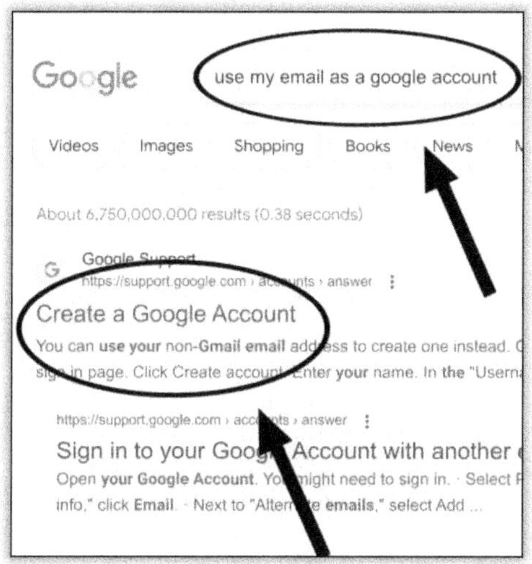


- Click on the link to Create a Google account.
- Scroll down and Choose "Use an existing email address."
- Click on "Google Account Sign In page" link.

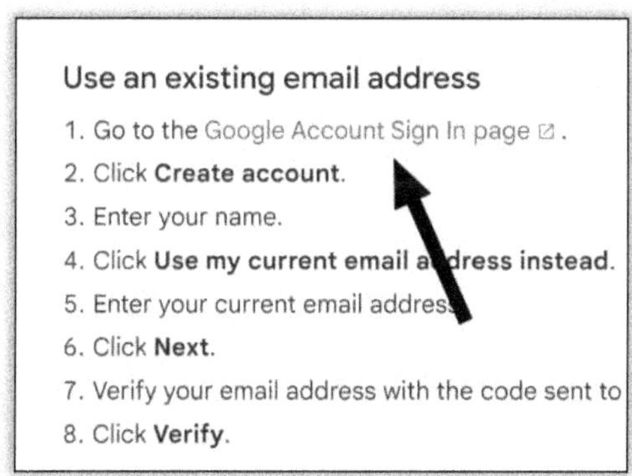

- You may have to click "Create an account" and give it your first name and birth date, but you will then see this screen where you can click "Use your existing email."

- On the next screen you will need to input the email address you want to connect to Google. They are going to email a code to that email address to verify it belongs to you.

- Enter the code from that email and verify your ownership of that email address. You will need to give Google some further account information such as a phone number (optional) and your birthdate and gender.

> *From Google:* "When you create a Google Account, we ask for some personal information. This info helps keep your account secure and makes our services more useful."

Then all you must do is agree to the Privacy and Terms and you are good to go! Your new Google account is set up using your existing email address or a shiny new, totally free Gmail address.

The importance of having a Google account will become clearer as we proceed with this book. A world of free and useful tools is now available to you. So, let's start "Using Google's Free S#!t" to grow our businesses to six figures and beyond!

We are going to dive right into the most important free tool that Google gives us, the Google Business Profile. This section will be a large part of the process of maximizing the return on all the Free S#!t, so we are going to spend some time going over each part of the Google Business Profile and explaining why each is important.

The following section is designed to be broken into seven daily parts. Keep this book near the computer you use to start your day, and aim for 5-10 minutes of reading and Google Business setup each day. In just seven days (less than two work weeks), you'll have a better understanding of Google Business Profile, and your profile will be properly set up to make Google love you and boost your business to the top of search results!

Let's get optimizing!!

1. What is the current email address you use for your company?
Is it branded to your company?
i.e.: robin@bbnb.org not bbnb.org@gmail.com.
List some of the free email addresses you currently use. (You know you do)

_______________________________________________

_______________________________________________

_______________________________________________

_______________________________________________

_______________________________________________

2. What are your 3 favorite free things?

_______________________________________________

_______________________________________________

_______________________________________________

_______________________________________________

_______________________________________________

_______________________________________________

3. If you didn't list 3 favorite free things from Google, shame on you, but what are your 3 favorite free things from Google?

______________________________

______________________________

______________________________

4. Have you ever seen the movie *The Sound of Music*? Who was your favorite character?

______________________________

______________________________

______________________________

______________________________

______________________________

______________________________

______________________________

______________________________

Remember, for updated information about this chapter, scan the QR code and watch the video for this chapter!!

Scan here to watch this chapters updated video

## George's story continued

Whew. Pretty tall order. Help my very non-technical friend out of a financial bind by helping him start and promote a new business?

This guy has no technical savvy, I don't even think he knows how to send a text much less post on Google or Facebook!!

Not only is his financial future at stake, but his children need him to support them and put food on their table. His kids and his ego really need this to work.

"Hey George, what's your email address?

"Success depends upon previous preparation, and without such preparation there is sure to be failure."

—Confucius

# SECTION II
# GOOGLE BUSINESS PROFILE

Setting Up Your Business to Maximize the Potential of
the Largest and Most Widely Used
Search Engine in the World.

# SETTING UP YOUR GOOGLE BUSINESS PROFILE

"The secret to getting ahead is getting started."
—Mark Twain

This is where the rubber meets the road. Is this just another cute business book or are you going to use it to increase your online presence and get more exposure?

Over the next seven days, we'll walk through setting up your Google Business Profile. Each exercise is designed to take just 10-15 minutes a day. I don't want you to feel overwhelmed by the time commitment or the tech side of things—that's why I've included detailed instructions, complete with pictures, arrows, and circles to guide you every step of the way.
Beyond setup, I'll also share valuable marketing and advertising insights from my 40+ years in the industry. Helping entrepreneurs grasp these concepts is what drives me,

and I love hearing when it clicks for you. It excites me to see entrepreneurs succeed when they apply these concepts. So, if you find this as exciting as I do, send me a text or email—I'd love to hear about your wins and challenges. Stay encouraged and expect great things. You're about to set your business up for success. You've got this. You *can* master the art of using Google's Free S#!t.

# DAY 1

1. Google the words "my business" (make sure you are logged into the right Google account). You should see this:

2. Click the "View profile" button. You will see "Your business on Google" and all the icons that will let you manage the parts of your Business Profile. The order they are in isn't always consistent, so I will put the picture of the icon with each description.

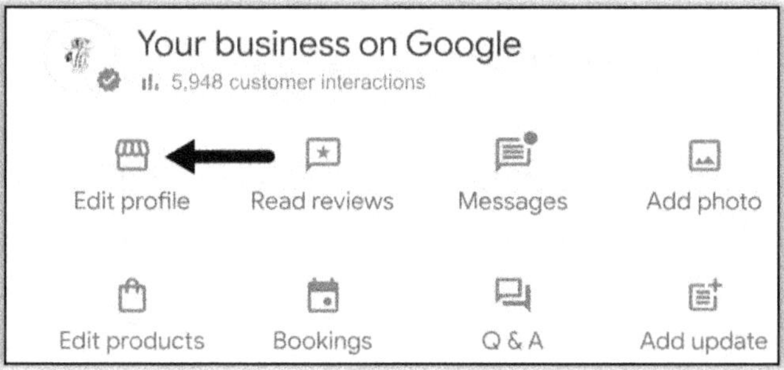

3. Click on the "Edit Profile" icon.
Here is where all the critical information goes. Google needs to make sure you are a valid business, and you are who, what and where you say you are.

This is your opportunity to make sure all the info Google will use to help their customers reach you is the freshest and most accurate info available. When their customer searches for a resolution to their problem, will Google be able to connect you easily?

Who do you think searches that term? When do you think they would be searching for such a product or service? Probably right when they need the service. Right when their problem occurs, and they need it fixed?

It is critical that you provide Google with all this information so they can be sure they are giving their customer the freshest, most accurate information available. And if you make Google look good, they want to showcase you as their best product. All Google cares about is that the customer gets good info and can make a good choice and feel confident about that.

Here is your first look at the business information section of the Google Business Profile. You see across the top the categories (About, Contact…) and you can scroll down the screen. You click the pencil icon that appears when you hover over each section to edit the info in that section.

Notice as you scroll down, the line under the category name at the top jumps from category to category. So, you can either just keep scrolling or click through the titles at the top to complete all the info. These are my best practices for setting up your Google Business Profile.

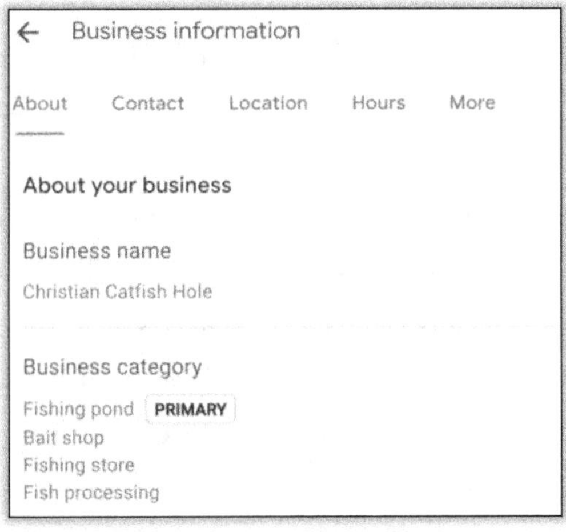


Google will customize this section based on what type of business they think you are. For instance, if you are a restaurant, they may offer to put your menu on your Google Business Page. If you are a salon, they will want to offer you a booking page to set appointments. They don't always get the category right, so some of the information in this section may not apply to your business.

## ABOUT YOUR BUSINESS

1. Business Name: Self-explanatory. This should be filled from when we originally set up the profile. This can be changed but you may need to reverify the profile if you change the name.

2. Business Category: You chose a category when you were setting up the profile. This is your chance to change that if there is a more fitting category for your business, or you can expand on it by adding additional categories that will help Google's customers to really know what you do. You can have up to 10 categories. Only the first one will show on your public Google listing.

3. Description: This is self-explanatory. However, as a veteran marketing/advertising professional, I would like to give my two cents about this. For years, we have heard that when promoting your business online, you should say your company name a lot and give long descriptions about what you do. This was especially true about your website because search engines scanned the text of the web page and indexed the words and phrases that were most used. Made sense, right? Google could very easily tell what your company did, and people would get the info they needed.

Current customers, however, will not read more than about two sentences before they become bored and possibly distracted. (Ooh shiny…) You need to be short and sweet here. Think of it this way. Google is going to use this description as a sort of SEO for your company. They verified this info with you, so they trust you to provide truthful information. Use this description to depict quickly and accurately what your company does, and possibly where you are located. Something like this:

---

> "Based in Dallas, GA iluvmygeek is your local computer repair shop that sells great laptops, desktops, and Apple® products. We also do in home and on-site service such as hanging your tv or rewiring your network. Slow computer? Broken laptop screen? Yeah, we fix that."

---

When describing your business, don't use industry terms and technical words. You are familiar with your industry and the buzz words and multiple interchangeable terms for the same item or process. Don't assume everyone else is familiar with those terms.

Think of how your prospective customer would be Googling what they need. Think about using some key words that someone that needs your product or service would be searching for when they need you.

This will help Google connect their customers with you and your business more easily. Helping the algorithm by using the same terms that the person that is searching will be using is another way to gain some algorithm love!

There is a lot to cover in the profile section. We will stop here for today and finish editing our profile over the next two days.

# DAY 1 WRAP UP:

1. Business Name – Check

2. Category: Brainstorm categories in your industry. Come up with 5 that complement your business.

________________________________________

________________________________________

________________________________________

________________________________________

________________________________________

Add these to your Google Business Profile categories.

3. Description: Read through your business description. Read it out loud, share it with a friend. Come up with a couple of compelling and inspiring sentences to add some flair.

________________________________________

________________________________________

________________________________________

________________________________________

________________________________________

________________________________________

________________________________________

## A FEW WORDS ABOUT YOUR CATEGORY (IES): (EXTRA CREDIT)

Google has allowed us to use up to 10 categories to describe our business. Think back to our discussion about being Google's product and consider that Google will use this information to connect us to their customer. Doesn't it make sense that helping Google to make connections between your business and other supporting or related businesses would go a long way? Take a few moments to really expand on what your business does.

Look at it from a user's perspective. What will they be googling when they need to find you? Maybe a customer with a tree issue might start by searching for a landscaper or yard person. Someone looking for a good cup of coffee might start their search by googling "café near me". Use these categories to give Google a better understanding of what you do and take a little creative license to give the algorithm something extra to think about when considering which cookie to give to the customer!

# DAY 2

1. Google the words "my business" (make sure you are logged into the right Google account). You should see this:

2. Click the "View profile" button and you will see all the icons that will let you manage the parts of your Business Profile.

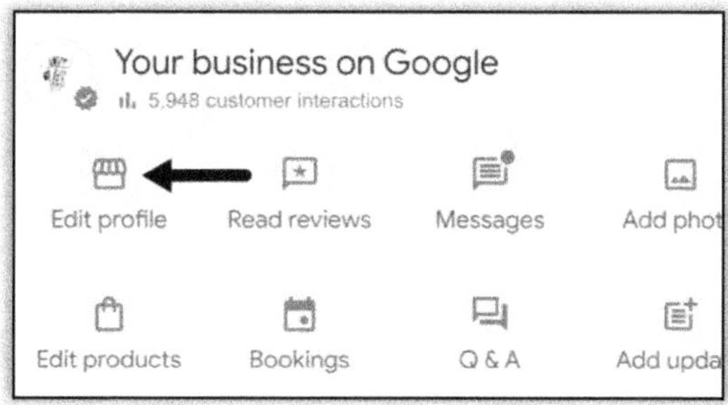


3. Click on the "Edit Profile" icon

We covered Name, Category and Description yesterday, let's jump back in with the next category: Opening date:

4. Opening date: This is the day you opened your business. If you have worked out of your home for 10 years and just opened a brick and mortar, this date would be 10 years ago. Google will increase your listing rank based on how long you have been in business. This also shows on your Business Listing as "xx years in business."

## CONTACT INFORMATION

5. Phone Number: Doesn't require any explanation. If you want people to reach you, make sure this is accurate. Also, if you are using an out-of-state area code for your business, consider getting a Google Voice number with a local area code. For more about Google Voice, see Chapter 7.

6. Chat: Google allows us to have chat button on our Business Profile. It used to be called "Messages" and it had its own icon like the "Edit Profile" does now, however Google incorporated the SMS (Messages) feature into the Edit Profile section. Here is where you can put in the phone number that you want your messages to go to when someone tries to message through your Google Business Profile. You can also turn this feature on or off, set up to use SMS or WhatsApp or both.

7. Website: This may seem obvious, but let's make sure we clarify a couple of things. Your website is one of the first chances your potential customer has to see and interact with your company. Having an updated website is extremely important when making that first impression. It doesn't cost a lot to have a great-looking site, and it costs even less to make sure it is updated regularly. Note about website: If you do not have a website yet, Google will hook you up with a free website that is simple to design and will work as a temp site or your permanent site, contact me for more information.

**LOCATION AND AREAS**

8. Business Location: If you have a location that customers need to be able to come to, this is where you list your company location so Google can give them directions to find you. If you work from home and you have clients come to pick up items or for meetings, you can list your home address; however, you will need to have permanent signage. If you don't display permanent signs that are easily seen, when Google sends a customer to find you the customer may become confused when they are directed into a residential area. Then, if there

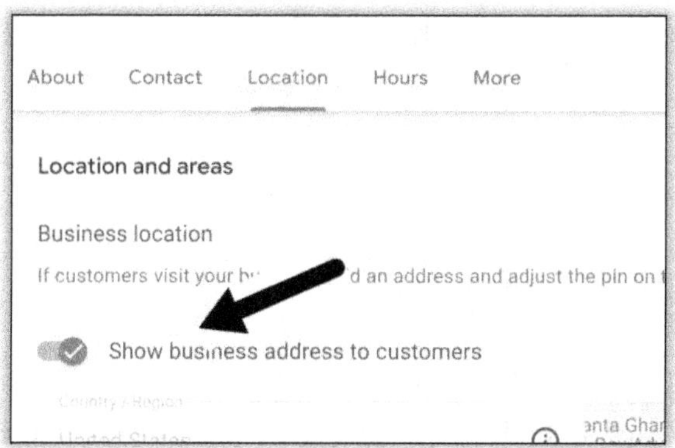

are no signs to indicate your business, it makes matters worse. In the absence of clear signs, you don't want your home address listed. Click the pencil in the location and areas section to turn off the "Show business address to customers" selector. Make sure you click "Save" in each section if you make changes.

If you aren't listing an address, you need to tell Google your service area. This is the area where Google will show your business to their customers. Understand that Google Business is a 'Local Search' product. Google's customers want local info when searching for a "near me" business, so limit the area you serve to a reasonable size. This way their customers searching for a product or service "near me" get only businesses that are close to them.

Have you ever stopped for the night in an unfamiliar town and searched for food nearby? With no knowledge of the streets or local spots, you rely on what's close. Google loves making that search quick and easy—helping customers find exactly what they need, fast. You are in a strange town for the first time, and you have no idea about streets or local areas. This is a great time to only get things that are close to you when searching. Google loves it when their customer finds the food they want quickly and close by.

We suggest you list the county you live/work in and the surrounding counties. Click in the blue "Search area" box and start typing. Make sure you select the right county. You can add as many here as necessary but remember Google Business

Profile is a 'Local Search' product. This means that if you say your service area is the whole United States, Google may lower your listing in the search results. Also, if you make your service area too large, you place your company in a larger and larger pool of companies that do what you do, thereby increasing your competition for eyeballs. (Think of Yellow Pages from years ago. You wouldn't expect to find Houston, TX listings in the Los Angeles area directory) If you are interested in more information about the Yellow Pages, just Google #yourgoogleguru and check out our YouTube channel.

To add additional areas, click the blue "Search area" box and start typing the next area you want to add. Once again, I recommend using the counties as they usually cover a larger area, and Google will include all the cities and zip codes in those areas, so you don't need to add them individually.

Ok, that is it for today, see you tomorrow!

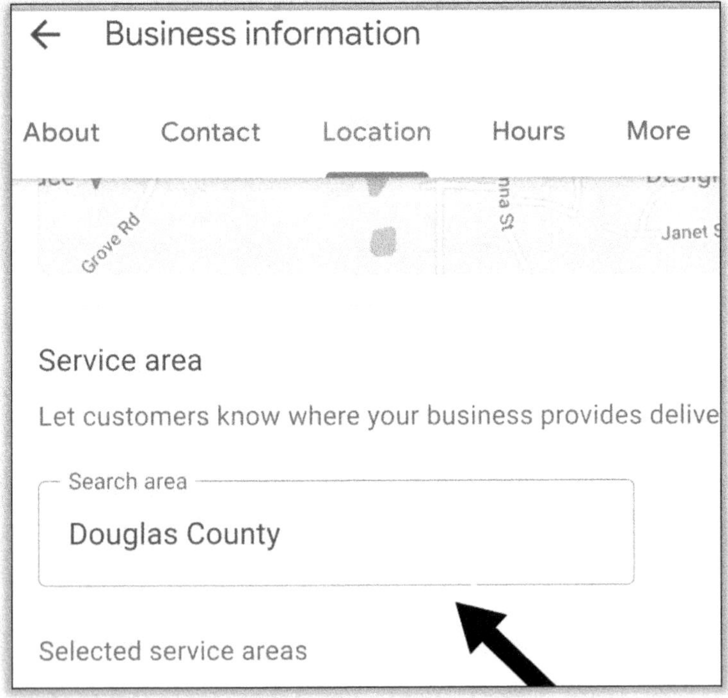

# DAY 2 WRAP UP:

4. Opening date – Checked. Remember to make sure this date shows how long you have been doing this type of work professionally. List the date when you first remember ever doing this type of work.

________________________________________

________________________________________

________________________________________

Contact Information:

5. Phone Number: How people perceive you is important. An out-of-state phone number can make a difference. Are you using an out-of-state phone number?

________________________________________

________________________________________

Are you aware you can get a local number for free?

________________________________________

________________________________________

6. Chat: Use Chat to change the number that messages will go to and choose between WhatsApp and SMS. List two reasons to use WhatsApp here.

________________________________________

________________________________________

________________________________________

7. Website: You MUST have a website branded to your business name. Does your domain name (www) reflect your company name?

________________________________________

________________________________________

Location and Areas:

8. Business Location:
If you are home-based business, you must choose your service area unless you have permanent signage. One example of a service area is to list the counties you want to be found in.
Are you a home-based business?

________________________________________

Are you showing your address on your Google Business Profile?

________________________________________

________________________________________

Thinking about a service area, list the county you are in and all the counties that touch your county.

________________________________________

________________________________________

________________________________________

________________________________________

# DAY 3

1. Google the words "my business" (make sure you are logged into the right Google account). You should see this:

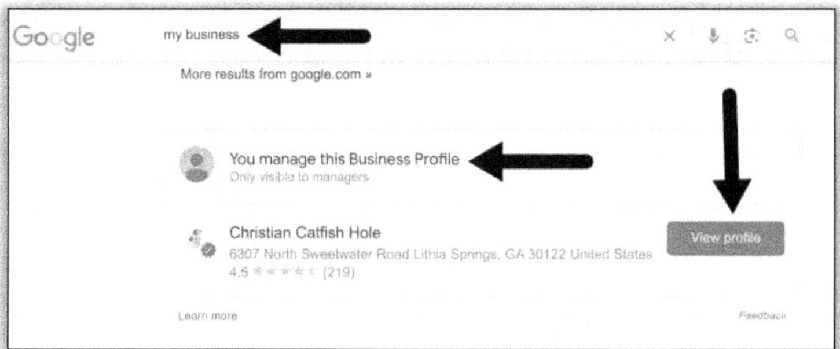


2. Click the "View profile" button and you will see all the icons that will let you manage the parts of your Business Profile.

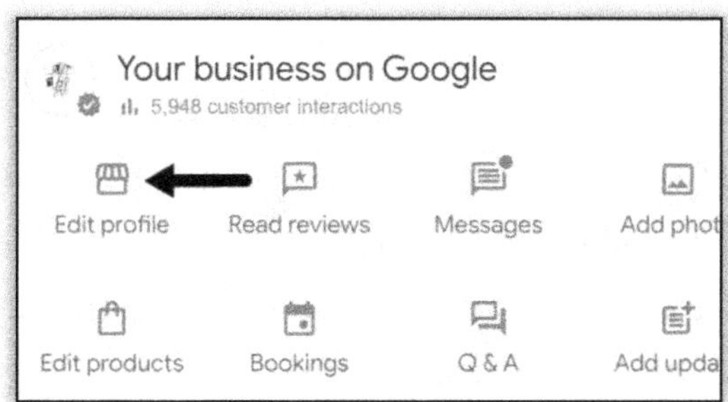


3. Click on the "Edit Profile" icon.

So far we have covered Name, Category, and Description on day one. On day two, we discussed Opening Date, Phone Number, Website, Business Location and Service Area. Now it is time to wrap up the Edit Profile section with Business Hours and More.

## BUSINESS HOURS

9. Hours: Self-explanatory. However, remember that if you list that you are open, and Google consistently gets feedback that you didn't answer the phone or weren't on site when your published hours say you are, you can set yourself up to be suspended. Google gives us access to edit this information very easily so we should remember to change our hours on our Google listing if we are going to be closed during normal, published operating hours.

10. Special Hours: Again, self-explanatory. But take a moment and fill this part out. Remember, Google is in the business of providing data to people who are searching. The more data available to them, the better. Google's algorithm can handle huge amounts of information, and by giving Google more information, you show that you care about their customer. By giving more info than your competitor provides, you position your listing to be placed higher on the search results.

**MORE**

11. From the Business: This is where you get to promote ownership identification.

12. Crowd: Other attributes that help customers make informed shopping decisions.

13. Planning: Do you have a calendar or scheduling page for them to book an appointment?

14.Service Options: Additional features Google offers based on your industry.

And that is it for the Edit Profile section. Remember if any of this information changes, get back here quickly and update your Google Business Profile to continue being the freshest cookie!

## DAY 3 WRAP UP:

Homework tonight: Google your industry and the city and state where you are located and check out some of the competition.

1. How many reviews do your closest competitors have?

________________________________________

________________________________________

________________________________________

2. Do they answer their reviews?

________________________________________

________________________________________

3. Read a couple of their reviews, read a couple of their answers. In your opinion, how did they do?

________________________________________

________________________________________

________________________________________

4. What category is each competitor using as their main category?

________________________________________

________________________________________

________________________________________

# DAY 4

1. Google the words "my business" (make sure you are logged into the right Google account). You should see this:

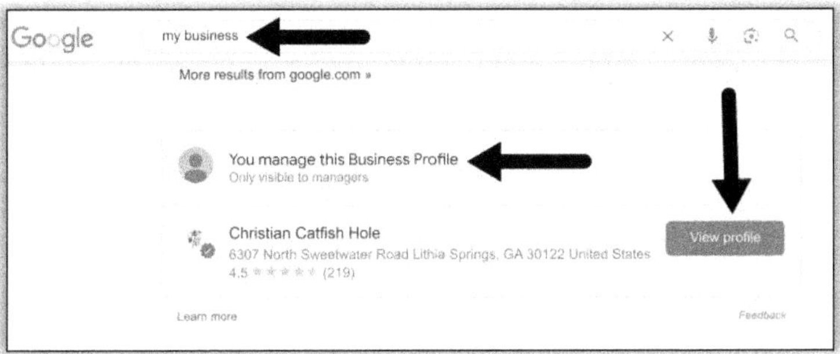


2. Click the "View profile" button and you will see all the icons that will let you manage the parts of your Business Profile.

3. Click on the "Read reviews" icon.

Here is an opportunity to really shine. Google and your future customers will read these reviews and most importantly, your responses! Follow these best practices when reading and responding to reviews:

- Respond to every review. This is to show Google that you appreciate their customers. Remember we need to show Google that their customer is top priority! If Google's customer took the time to leave the review, you darn sure better thank them for it.

- Don't use a canned response. Review readers will see right through that, and they will think you are either deceptive or lazy. Take two minutes to type a unique sentence or two thanking each customer for the review.

- Make a comment about the service or repair that your company performed. This shows you did something for the reviewer and makes the response more personal.

# DAY 4 WRAP UP:

Go check out any reviews you may have already received.

1. Do you have any reviews that you need to respond to?

________________________________________

________________________________________

2. Do you have any negative reviews?

________________________________________

________________________________________

3. Check out your previous responses to customer reviews. Now that you are learning more about Google Business Profile, how could you update some of your older responses that may not seem professional?

________________________________________

________________________________________

________________________________________

4. Check out any messages you may have already gotten. Familiarize yourself with where to look and how to respond. Change your auto response message.

5. Write 3 questions that you get asked the most about your business.

______________________________________________________

______________________________________________________

______________________________________________________

______________________________________________________

______________________________________________________

______________________________________________________

6. Now take those 3 questions and create 3 FAQs for your Google Business Profile!

______________________________________________________

______________________________________________________

______________________________________________________

______________________________________________________

______________________________________________________

______________________________________________________

______________________________________________________

______________________________________________________

______________________________________________________

______________________________________________________

______________________________________________________

# DAY 5

1. Google the words "my business" (make sure you are logged into the right Google account) you should see this:

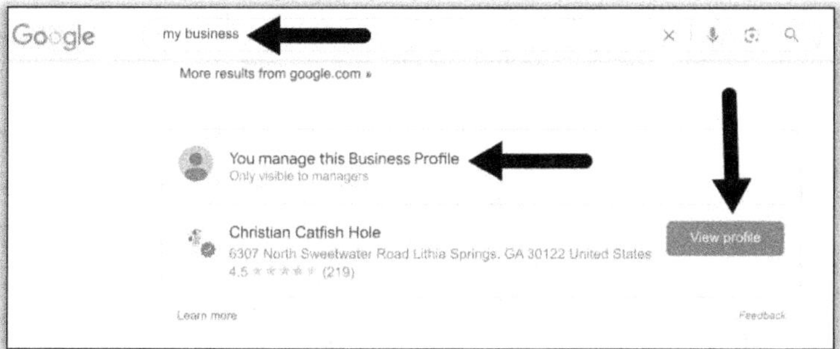


2. Click the "View Profile" button and you will see all the icons that will let you manage the parts of your Business Profile.

3. Click the "Photos" icon.

The first thing you see here is "Photos and videos." This is the common, general photo area where Google allows your business to post photos. These are public photos that customers can see as well as where customers can post photos of your business.

The first two pictures in the list are labeled "Change cover" and "Change logo". This is where you will upload your cover photo and your company logo to be used on your Google Business Profile. Your logo will be displayed in a round frame, so I recommend saving your logo on a white background that is about two times larger than the image of your logo. Save this as "Framed Logo." Then when you use this file where it is going to be displayed in a round frame, it won't cut off your logo to make it fit.

These two pictures are both 500px by 500px, but notice I made the one on the right much smaller but kept the background the same. This presents a more professional image, and let's face it, customers are factoring all this stuff into their decision about which company to work with. Details people!!

Your Cover Photo is the one that will be displayed on your Google Business Profile. This is your chance to catch people's attention and hopefully get them to want to look at your company. Your cover photo isn't your logo, that appears elsewhere. The other photos showing are the map of your location (or a shaded service area if you don't have an address displayed) and a photo of the outside of your location. (Again, this whole section would show the service area if you don't have a physical location to display)

Besides your Cover Photo and your Company Logo, Photos are a big part of making Google's Algorithm (AI) happy. But remember this is not social media. The pictures you will post here are less about showing people your business as much as they are about getting activity. Google wants to share interesting businesses with their customers.

Getting more photo views will generate more activity in your profile. More activity means you may possibly be more interesting than your competitor. AI likes that and raises you higher on the search results list. This is one small part of the overall formula that gets a higher organic ranking on Google.

Posting pictures from time to time will help keep your photo views up, and you can post photos when doing updates to your Google Business Profile, which is great. Getting customers to post a picture with a review is amazing. But another way to use Photos to get Google's algorithm to notice you is to post 20—25 random pictures all at once. Yes, I said random. Yes, I said all at once. Yes, I said 20—25 pictures, and no, the content of these pictures is less important than just getting them posted. Here is the madness behind the method:

When you post that many pictures all at once, some people will go and look at them. This will show up to Google (Al) as activity. This will help you grow from 50–60 daily photo views to 400 (and that is if you only get 20 views of each of your 20 photos). This registers with Al and he wants to know what all that activity is about. You seem more interesting, and you are elevated in the search results. This isn't something you want to do regularly. Instead, do this twice a month to greatly increase your chances of ranking higher in organic Google search results.

4. Click the "Performance" icon

Here is where you get to see how your hard work is paying off. You can access this information through the "Performance" icon or if you look where it says "Your Business on Google" you see how many interactions your business has had in the last month. You can click on the number of interactions to access the performance information as well.

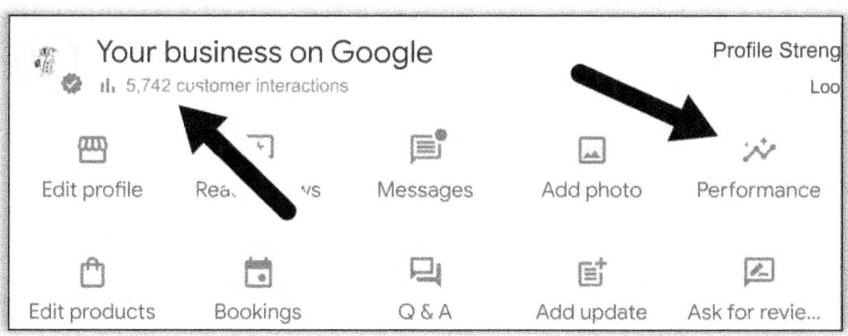


Performance is an "analytics junior" for your Google Business Profile page. It shows you how many views your profile has gotten and what people did once they saw your profile page. When you click on the icon you will get a different page with some categories across the top of the window.

At the very top is a date range selection box. You can change the date range by clicking on the dates and selecting the beginning month and then the ending month. The default is the last six months.

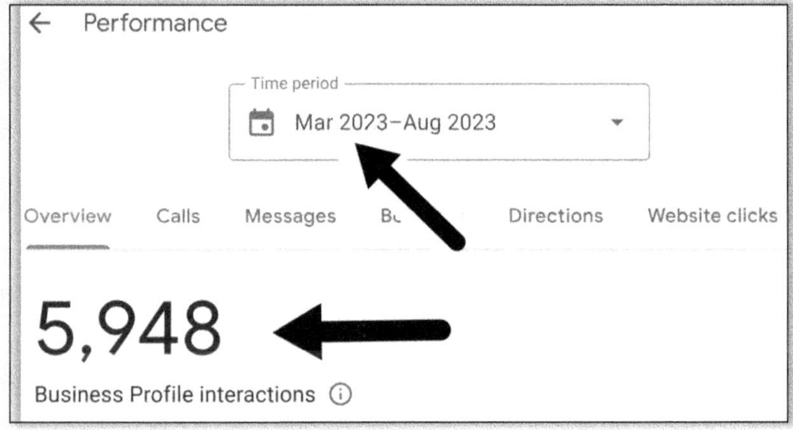


The first category is Overview. Here you will see how many interactions your page had during the selected period. As you scroll down, you have a lot of data about how Google's customers are interacting with your profile page.

If you have products listed, it shows how customers interacted with them; it also shows how many people viewed your profile, and further breaks that down by device and which platform they interacted with (Search or Maps). In the right column, you see the key words used to find your profile. All this information has value in tracking your history and directing your future business decisions.

Across the top the next category is Menu. (If you use an online menu) then Calls and Messages. Remember this information is strictly about the interactions from your Google Business Profile. Next is Bookings, again somewhat industry related (hair, nail salons, barber shops etc.)

The next two categories are Directions and Website Clicks. This is great information to help you further understand your potential client and their buying process. Knowing how many customers clicked on your website link and comparing that to your website analytics can show the power of Google and what percentage of website visits you are getting directly from your Google Business Profile page.

Ok, let's close the Performance window by clicking the arrow at the top left of this window.

5. Click the "Advertise" icon.

This will immediately take you to https://ads.google.com and initiate a new Google ad campaign. This book is not about Google's paid offerings. There are many appropriate times and promotions that greatly benefit from Google ads. That is for a different book.

If you are a restaurant or lounge, or if Google thinks you may fit into a similar category, you may see Edit Menu and/or

Pickup and Delivery icons. These are very self-explanatory, and Google will even give you performance metrics for your top performing products!

If you aren't in the above categories, you will most likely see an "Edit Products" icon and an "Edit Services" icon.

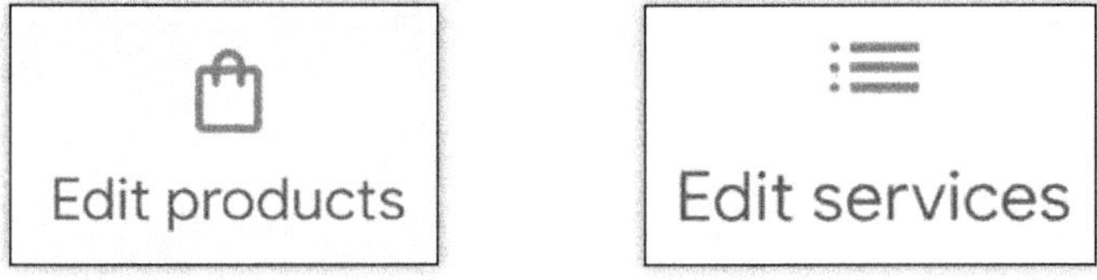


The Edit Products and Edit Services categories are exactly what they say. You can list as many products and services on your Google Business Profile as you want. Google will then put those on your listing when you match a Google search. In the description of your products and services, you can list the keywords and phrases that people would be searching for so Google can match you up. This is your main keyword area in your Google Business Profile. You can get unbelievably detailed in your SEO keywords by using the descriptions about your specific products.

Keywords and SEO are two of the most feared words to businesses needing to get found online. Most marketing companies don't really explain them clearly, and this leads to some of that fear. Keywords are simply the words that will help Google match you up with their customer. Keywords help Google determine what you do and then they index those words so they can find them easily. When their customer (the person searching) searches for those "key" words (or terms) the algorithm will place the company with the proper keywords in the search results.

## DAY 5 WRAP UP:

1. Look at your photos on your Google Business Profile page: How many do you have?

________________________________________

________________________________________

2. Locate some good pics you could upload to your profile page. Write down where they are located on your computer or cloud storage.

________________________________________

________________________________________

3. Set your calendar to remind you to upload 20—25 pictures to your Google page every 2 weeks for the next 3 months

Go to your performance icon. Scroll through the sections.

4. What are the top 3 keywords being used to find your business?

________________________________________

________________________________________

________________________________________

This is important as you can use this information to help drive your advertising focus.

My sign company found that 80% of Google Business Profile visits came from searches related to yard signs. Now, we have a dedicated webpage to help customers with their yard sign needs!

# DAY 6

Click on the Q&A icon.

This is exactly what it sounds like. Here you can set up a simple Question and Answer page for your business.

Having a Q&A page serves several important purposes:

1. Customer Engagement: This lets customers ask questions and get answers directly on your Google Business page.
2. Info Sharing: Here you can share important information about your business. All the common questions about products, services location and more.
3. Customer Interaction: This lets you build trust and rapport with your customers. Answer quickly and provide good information!
4. SEO: The information in the questions and answers will help Google and provide your business with better visibility.
5. Customer Perspective: This gives you a different angle on your business. Learning how customers see your company.

To get the most out of your Q&A page, monitor it regularly and respond quickly. This will help enhance your online presence and show Google you are on top of customer issues and satisfaction.

Click on the Add Update icon.

Clicking the Add Update icon will allow you to post to your Google Business page. There are a lot of reasons you want to post to your Google Business profile. The 3 types of posts are:

1. Add Update
2. Add Event
3. Add Offer

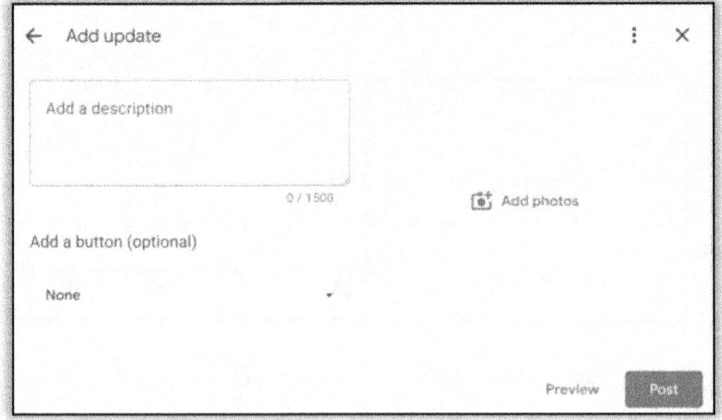

All of these will create more activity and gain interest from Al. The Event and Offer options are great for just what they say. If you have an upcoming event, you can post it here. If you have some great special offer you want to promote, post it here.

But no matter what you have going on with your business, you want to post on your Google Business Profile at least three times a week!

Now I cannot tell you exactly what time of day or what day of the week you should post, but a minimum of three times a week is non-negotiable. This is a remarkably simple way to gain a reputation with Al. Just the activity of posting the info as well as any interaction of people reading it will spike your algorithmic activity and gain more recognition.

The "Add Update" option has fewer requirements to post than the Offer or Event Posts. Picture. Text. Button. That's it. And you don't even really have to put a picture, but people like pictures and it may gain you a bit more activity.

Back in the old days you could count on the general store owner to show up early, put on his leather apron and hat, prop open his front door and then sweep the front porch. This showed the townsfolk that his store was open for business. Nowadays, we have that ubiquitous flashing sign in our window: "O. P. E. N." This calls to our customers during business hours. But Al cannot see our open sign. How do we tell Al we are open and available for him to send his customers to us?

What you are doing by posting three times a week is showing Google that you are available and open for business. They are not walking down Main Street and they can't see your front door, but when you post an update regularly that lets the algorithm know that you are open for business. By default, this will cause you to get more activity on your profile.

This isn't social media and does not require it to be a specific topic or even a beautiful picture. Just the post itself will gain the activity and that will help boost your business in the search rankings. You can post about anything; it doesn't even have to be related to your business, and it doesn't really even need to be pertinent to some current event. It just needs to be done so that you get the activity, and the algorithm notices you. The way we make sure we are always posting is to use another of Google's free products which is their Calendar. We remind ourselves to post by putting the event on our calendar and that way it is not forgotten.

## DAY 6 WRAP UP:

Calls: Turn call history on. It really doesn't have a huge amount of value these days due to the calling history of our cell phone carriers. Also, Google Voice has an amazing call history and will keep your calling, texting and voicemail history forever.

1. Have you heard of Google Voice?

___

___

___

Q&A: Start creating some Q&A topics so you can show Google that you are giving their customers as much information as possible.

2. Write 3 common questions you are going to use for Q&A page.

___

___

___

___

Add Update (Event/Offer): Add an update. Take a picture of your desk and post it with a short comment about what a great day you are looking forward to (or just finished) and put a link to your website.

3. List 3 topics you can use for future updates.

________________________________________

4. Set up your Google Calendar to remind yourself to post on your Google Business Profile.

5. Look at some of your competitors' Google pages and see what they are posting.

6. Do your competitors have any FAQs.

# DAY 7:

Click on the Ask for Reviews icon.

When you click the Ask for Reviews icon, you are going to get a link. This link is worth its weight in gold. Back in the day, we used to have to struggle with how to link our customers to our review button. There was no direct way to send them to the exact spot without having to give some lengthy directions with pictures and arrows! We found some sneaky ways such as pulling up the review box in Google Maps and copying the URL. This sometimes worked and sometimes didn't.

But all that is behind us now. The Ask for Reviews button reveals a cool link that when Google's customer clicks on it, they are taken to the exact spot where they need to be to leave the review. If they are logged into their Google account when they click the link, they will get the window showing the five blank stars and the area to write a review.

There is also a place to upload a photo, not necessary, but always appreciated.

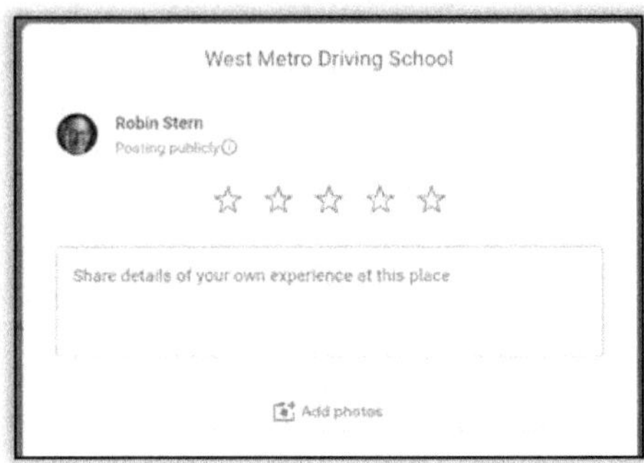


This link has revolutionized getting reviews as there is nothing in the way of the users being able to find where they need to be to show some love for the amazing service (or product) that you provided them. In the next chapter, I will show you a way to use this link to get even more algorithm love than a five-star review can generate on its own.

# DAY 7 WRAP UP:

1. Think about the last time you asked for a review.

2. Were you uncomfortable asking?

3. If you asked, did the customer leave the review?

4. Do you have a simple method to help your customer find where to leave the review?

5. How many reviews do you have?

6. Check out some similar businesses to yours. How many reviews do some of your closest competitors have?

George's story continued

First of all, George, I need you to learn some of this new computer/website stuff and I need you to make sure you do as I say.

"You got it Rockstar! I don't have any idea what I am doing, but I can follow instructions. I will do EXACTLY what you tell me to."

Ok. We need to setup a website and a Google page, maybe a Facebook page as well, and you need to work on posting and updating regularly.
"Ok show me what to do" George replied.

So we set up a simple website with "before and after" pictures of his jobs, and a picture of George with all his kids sitting around him. Then we set up his Google My Business page.

Remember, for updated information about this chapter, scan the QR code and watch the video for this chapter!!

# SECTION III
# THE REST OF THE STORY

How much money *can* you save?

# REVIEW US ON GOOGLE

> "Twice and thrice over, as they say, good is it to repeat and review what is good."
>
> — Plato

The Ask for Reviews icon is very important in getting reviews, but it has a much greater value. Sending that link to a satisfied customer after you are finished with him/her will increase your positive reviews. We have seen clients getting upwards of 90% of the reviews that they ask for when they follow our system.

However, this also creates a problem with getting traffic to our website. Google is so good at getting our information in front of our potential customers that sometimes that may create a lack of customer traffic to our website. This can be an issue since traffic to our website is a portion of what the algorithm uses to rank us on search results!

What if you could drive more traffic to your website and get more page ranking, as well as get that coveted review that you so richly deserve? Let's look at this.

When you click on Ask for Reviews you have a link that will open the review box so the user can leave you a review. You have a website that needs more traffic. What if you were to put the link for the review as a button on your website?

Then when you send your customer a text after the sale, you send a link to your website and direct them to click the button that says, "Leave us a review on Google."

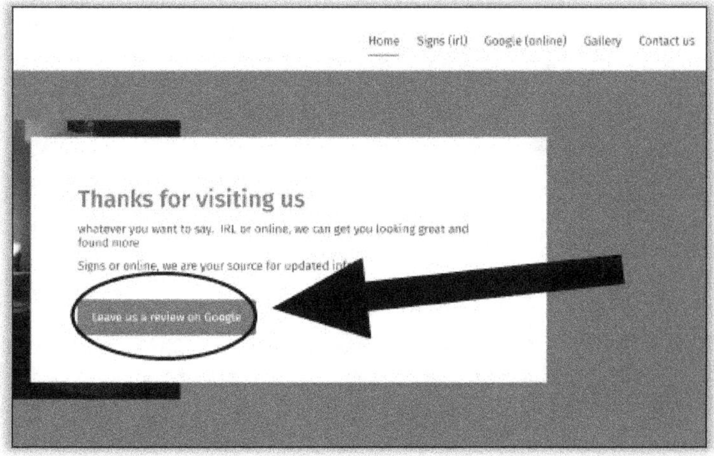

This is a fantastic way to not only get more traffic to your website, but you could also use this opportunity to get more information in front of what is already a satisfied customer. This could lead to further business with this customer. Maybe you fixed their computer, but they don't know that your company installs tv's on the wall, can fix their wireless network, and sells great refurbished computers! This would be a chance to get that information in front of these people.

# RECAP:

- "Ask for reviews" link from Google Business Profile

- Place review link on your website as a button.

- Have a text script to request the review with link and directions for how to find and click the button to leave the review.

Here is an example text script:

"Thanks for choosing us for your needs. Would you please click this link to our website where you will see a "Review us on Google" button. Thanks!"

This is always sent as a text as soon as possible to capitalize on the good feelings that your customer is experiencing. This will allow them to feel even better as they take the time to review you.

Again, sending this text as quickly as possible is imperative to give yourself the best chance to get the review.

1. Can you edit your website and add a button? Jot down the website login info, including the username and password.

2. If you answered no, can you get your web designer to add a button to your website?

3. Jot down the person who works on your website. Include their phone number and email address.

4. Do you know where/ how to get your review link from your Google Business Profile? List the steps to log into your Google Business Profile:

_______________________________________________

_______________________________________________

_______________________________________________

_______________________________________________

_______________________________________________

_______________________________________________

_______________________________________________

5. List three places on your website where you could put a button that customers could click to leave you a review.

_______________________________________________

_______________________________________________

_______________________________________________

_______________________________________________

_______________________________________________

_______________________________________________

6. Knowing your average customer, how can you word your request for a review to inspire more people to take the time and leave you a review? Take 5 minutes and jot down some things you would like to try. Include any words or phrases that got positive results in the past.

Remember, for updated information about this chapter, scan the QR code and watch the video for this chapter!!

George's story continued

Getting George to use a computer was impossible, but when I explained to him the importance of being able to post updates, answer reviews and keep his Google Business Profile active with photos, he was willing to try.
"Rockstar, I know I have to do what you tell me. If you say I need to ask for reviews, answer the ones I get and post updates and pictures, then that is what I will do."

Ok George. Just keep doing that.

And he did. Not only did he do the technical things I told him to do, but he also added a touch of his own personality and southern charm. He built relationships with every customer; he saved their numbers in his phone so if they called again, he could answer the call with their name. He began to master the review request, and his 5-star reviews started taking off.

He started to get more calls, and every one of them helped him get closer to financial freedom for him and his children.

## USING BRANDED EMAIL VS. GMAIL

> "Products are made in a factory, but brands are created in the mind."
>
> — Walter Landor

There is always discussion about using a Gmail email address as a business email. The ultimate answer is: if you want to be seen by the public (and your potential customers) as an actual business, you need an email address that is branded to your company. So, yourcompany@gmail.com, is not as good as yourname@yourcompany.com.

We discussed this in a previous chapter, but for the purpose of setting up a Google Business Profile, you can use a Gmail email address if you want since this email address is rarely seen by the public, so your brand image won't be impacted. But please, if you do have a public facing email address, truly consider investing in a branded email address that reflects your business or what you do, not a Gmail, Yahoo, MSN or God forbid, an AOL email address!

Once you decide to use your company name as your email address, you have some other things to consider.

- Do you own your company name as a .com domain name?
- If not, is the .com available?
- Where do you set up your email?
- Does it cost to have a branded email address and if so, how much?

These are only some of the questions we need to ask. If you want to check out some available .com domain names, there are many places to register them online. Google the words "domain name registrar" and check out some of the ones you get there. Just make sure to shop prices and read a few reviews. Different companies have vastly different pricing plans. If you want some guidance, I would be glad to make some recommendations. Google has done domain registration before, however they may be selling this part of their organization in the near future.

Once you find out if your dot com is available, you can purchase it and then you will need to decide how you want to set up your email. You can use Google's email system; they call it Google Workspace.

There is a charge for each email address, but you get access to Google's powerful email system as well as all the other features that come with Google Workspace.

There are other options. You can get unlimited email addresses for as low as $2.00 per month with some website/email hosting companies. Please contact someone you trust that can assist you or reach out to our organization for further guidance.

If you cannot afford to pay for your email, just realize that you could be hampering your forward momentum. People see Gmail, Yahoo, MSN or AOL email addresses and they consider you a hobbyist, not a professional.

Matt Pearce from the University of Georgia Small Business Development Center (ret) stated in a presentation to the Paulding Chamber of Commerce that for every business that failed in 2020, there were 3 commonalities:

1. Not enough capital.

2. Not enough marketing budget. (And when they ran short of capital, the marketing budget was the first to get axed.)

3. An unbranded email address.

1. Where is your website hosted? (GoDaddy, Wix, etc.)

____________________________________________

____________________________________________

____________________________________________

____________________________________________

2. Do you pay for your email address?

____________________________________________

____________________________________________

____________________________________________

3. If yes, how much do you pay a month?

____________________________________________

____________________________________________

____________________________________________

____________________________________________

____________________________________________

4. Look up 3 or 4 of your competitor's websites. Check out their contact information. Are they using a free, unbranded email address?

____________________________________________

____________________________________________

____________________________________________

____________________________________________

____________________________________________

5. Whether or not your competitors are using free email addresses, how does that make you feel about their companies?

____________________________________________

____________________________________________

____________________________________________

____________________________________________

____________________________________________

Remember, for updated information about this chapter, scan the QR code and watch the video for this chapter!!

## George's story continued

"Rockstar, I cannot believe how many calls I am getting. I can't keep up with it all. I have professional sports stars that keep calling me to come remove all the stuff other companies either won't take or want way too much to haul it away."

George was really embracing the system. He trusted me completely, so he just did what he was told. Asking for each review, sometimes even helping the customer figure out where to go so they could leave the review. Responding to each review in a timely manner. Taking lots of pictures so he could have content to post on social media and of course on his Google Profile. He is the poster child for my system. (Maybe that is why his story is in my book!)

His branding was really starting to catch on. From the amazing company name to his branded email and marketing materials, his company was starting to become the first call for anyone needing junk removal.

And then the calls from his competitors started….

# MORE OF GOOGLE'S FREE S#!T

> "The price of anything is the amount of life you exchange for it. "
>
> — Henry David Thoreau

When it comes to some of the other cost-saving features that we use with Google, we have calculated that in the year 2022 our computer company saved $20,000 by using Google's free products and services. Here's how we calculated that estimate.

First, by optimizing Google Business Profile, which we estimate replaces about $1,500 a month worth of advertising. Second, add in the value Google Voice, which we used as our main phone number for our company. We figure this saves us about $250 a month for phone service. Those two items alone come to more than $21,000 a year in savings!

We also use Gmail for some of our in-house emails, which gives us access to Google Drive, Google Docs, Google Slides, Google Sheets, Google Forms, and the rest. We're going to explain some of these products one at a time so you can get a better idea of what each does and what the value is to our company.

There is not enough room in this book to include all the additional free things from Google. You can see all things Google at https://about.google, and it is a long list. In this book we share the free things that we use. Tried and true products and services that we have used in the daily operations of our companies for more than 10 years. That is not to say these things can't change. The tech industry is constantly changing. If they do, we will update this book and create new videos for the updated information. (You can also join our Facebook group, which we mention at the end of this book)

One of the most cost-saving products we use every day is

## GOOGLE VOICE

Voice is a communication tool from Google. There are many features that make it so versatile, and we will go over some of those here. First, head to https://voice.google.com and sign into your Google account (see chapter 3 for a discussion about your Google account)

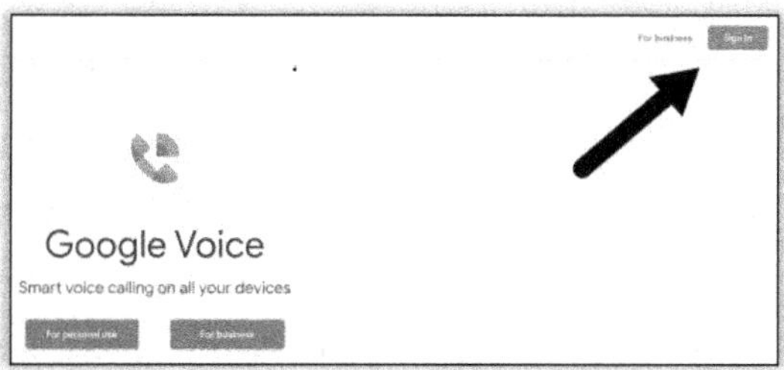


Once you are signed in, you will see the welcome screen where you can read the terms and conditions and you will click "Continue."

On the next page, you get to choose your phone number. This can be any number in the United States; you don't have to select a local number. Just start typing in the location or the area code you want and Google Voice will display a list of available numbers. If no numbers display, that means there are no numbers available for that location or area code.

> "When we were planning to move to Georgia from Florida, we picked a phone number local to the town we were moving to. This is useful, as you can change the number at any time, making this a truly universal communication tool."

After choosing your number, you need to tell it which phone you want to send the calls to., Remember this is a virtual number. That means there will be no actual phone associated with the number. It also means you get to pick what phone you want to send the Google Voice calls to. You can have this phone number ring up to 6 different phones at the same time, and the calls can be transferred between those 6 different phones. These can be landline or cell phone numbers, and once they are connected to your Google Voice number, you will be able to answer your calls on any of the 6 numbers you attached to this account. For more detailed help, Google has great online resources, or you can reach out to me for any issue you may be experiencing.

## GOOGLE FORMS

Google Forms is a form-building platform that you can use to create any type of form you may need. Whether this is for collecting data or for gathering names for Christmas gift-giving lists, you can use a Google Form to capture information.

In our computer business we use it as a method of checking computers in for repair. We have a large touch screen monitor in our lobby and when someone brings their computer in for a repair, we ask them to fill in the Google Form.

We have been using this for more than 5 years to the tune of 2000 plus forms. This allows us to capture their name, their address, and all their contact information and also the computer password, the problem they're experiencing with their computer, and check boxes for them to select so they can agree to our terms and conditions, our data retrieval policy, and our abandoned property policy.

After they've filled in this form and agreed to the policies and terms, when they hit "Submit", it puts it into a list where we can now print a copy of that form and use it to track the repair through the shop. During our first four years getting started up and set up, this was a very cost-effective way for us to not only keep track of each repair but also the date of repair. If someone came back to us and said "you fixed my computer 6 months ago", we can go back to those forms and find the one from their repair and we know exactly what date it was repaired and exactly what we did.

We also use forms to remember the sale date when we sell somebody a computer. We offer a one-year warranty on all computers that we sell so if you have any problems, you bring them right back to us and we will fix them. This is easily tracked when someone purchases a computer. We ask them to go to the touch screen monitor and fill in the form for the purchase. This not only gives us all their contact information, but also the date of the sale so that we know going forward when their 1-year warranty expires.

Google Forms has dozens of templates that you can start with, or you can start from a blank form. It's very easy to drag your fields into the form and then customize it the way you need it. This allows you to capture all sorts of data and it also allows you to export it as a CSV file so you can import that information into other platforms or generate mailing lists or whatever you need for your intended purposes.

Again, this is a product that costs our company absolutely nothing and we have been using it to gather and keep track of our customer info for the last 4 years.

## GOOGLE CALENDAR

We use Google Calendar for all our scheduling in the office. We schedule employees' hours, all our on-site customer repairs, vacations, etc. The ability to track that across any platform and the ease with which Google Calendar integrates with all other platforms has made this an amazingly versatile tool for our business.

The features of Google Calendar never seem to end. I was on a networking call recently and someone mentioned that you could attach files to your appointments in Google Calendar. My mind was blown! I had no idea all my documents, photos, videos, spreadsheets, whatever was necessary for that meeting could be attached right to the calendar entry and then when the calendar popped up, all you had to do was click on each item. I didn't have to search to find it; it wasn't like it took any amount of time. It was saved in Calendar and was readily available to me when I pulled up that appointment.

This is an incredible tool for any company that is scheduling salespeople for in-home visits, or possibly any sort of repair company that has assistants scheduling the calls or the repairs for their people. You could attach any documentation that the salesperson or repair person would need to the actual calendar entry, and when they got to the customer, they would have all of the documentation they need right at their fingertips.

It's amazing to me how every day I learn more about different services and products from Google. I haven't found any app or online platform that has a problem integrating Google Calendar. From Calendly to the CRM that I use to our QuickBooks, all these are very easy to integrate with Google Calendar and make life an absolute breeze.

Being able to view multiple calendars on one device also helps. For instance, my personal Google calendar, my non-profit Google calendar and the computer shop's Google calendar all show when I open the Calendar app on my phone. When I open the calendar, I see different colors for each of the calendars. For instance, my personal is purple, my non-profit entries are light blue, and all the stuff the guys put in from the shop is gold.

My CRM is now interfaced to my Google calendar to make sure I don't overbook myself, and the entries that people make from my CRM when they book an appointment with me show up on my Google calendar in red. This is a quick way to be able to determine what's on your schedule, where you need to be, and which hat you need to be wearing.

As an entrepreneur, teacher, and executive director of a non-profit, I have a lot of different responsibilities and my Google calendars keep them all separate. Once again I can attach any supporting documents, videos or audio clips, whatever it may be, to the actual appointment entry and they're right there with me when I open my calendar.

Google calendar allows multiple people to be logged in at the same time, and you can be logged into it from multiple locations simultaneously. This keeps everyone on the same page and allows us to see where everyone is at any point during the day.

## GOOGLE SHEETS / DOCS / SLIDES

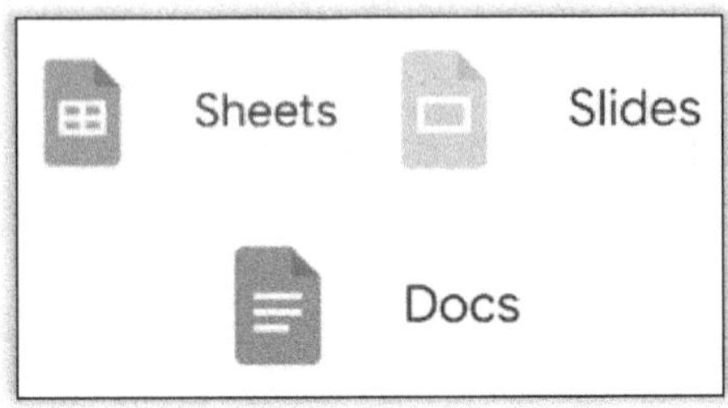


When it comes to using all the Google office products which mirror Microsoft Office, you access them the same way as all the other Google services and products.

After signing into your Google account with your browser, click the nine dots in the upper right-hand corner and you

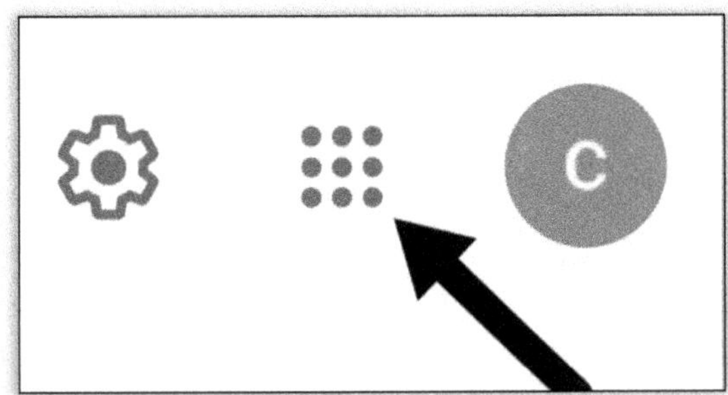


will see a list of all the available applications and services from Google. If you click on Docs, Slides, or Sheets, you will notice that each of them is similar in look and feel to its Microsoft counterpart.

- Google Sheets is an amazing spreadsheet program that is laid out to look and feel very much like Microsoft Excel

- Google Slides is a slide presentation program that is laid out to look and feel very much like Microsoft PowerPoint

- Google Docs is laid out to look very much like Microsoft Word.

Now this isn't to say these are direct replacements for the Microsoft products. However, they are free, and you can access them from any internet-connected device. If you are using Google Drive to save all your documents, slide presentations and spreadsheets, you can now access all of the content that you have created with these three programs from anywhere in the world that has an internet connection.

## GOOGLE DOCS

One way to access Docs is to click the 9 dots in the top right corner of your web browser. Another is to navigate to the website: https://docs.google.com.

Here you will see:

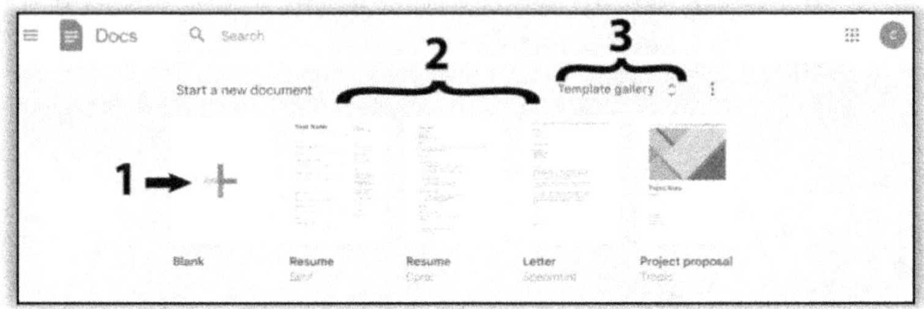


1. A large + sign for you to create a blank document and format it yourself.
2. A short list of available templates across the top of the screen. There are many premade templates to choose from, and you can access all the available samples here.
3. The "Template Gallery" at the top right. There are many premade templates to choose from. Below you will see the most recent documents you have previously created.

Let's start with a blank doc so we can look at the Google Docs interface.

Here we can see what is somewhat like a Microsoft Word screen.

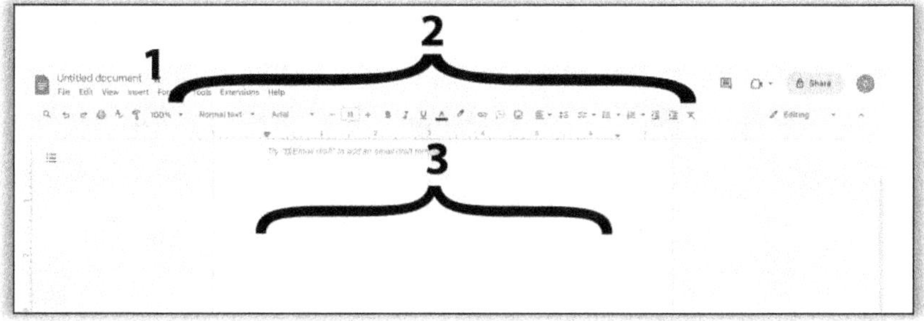


1.File name
2. Format and features ribbon
3. Working area – your blank sheet of paper

I have spent the last 15 years going back and forth between Microsoft and Google when it comes to document editing. I'm leaning more towards putting everything on Google as the internet connections get better and we are more able to access these programs and files from handheld devices.

Some of the benefits of working on a web-based document editor are:

- You never have to worry about saving your file. It constantly saves the file after almost every keystroke.

- You can access your files, edit them, and even print or present them all from the cloud. This means no more lost files, no more "where is that thumb drive" and no more "can't read disk" errors from your laptop. Google cloud-based programs save your files to your Google Drive by default.

- Cost. We did mention all the stuff we talk about in this book is completely free, right?

One of the limitations we have found with using Google Docs stems from migrating documents you created in Microsoft programs over to the Google-based programs. Some of the specific formatting features in Microsoft Word do not translate well into Google. There can be problems from time to time when you're trying to open documents that were created in Microsoft Word. However, these minor irritations are far outweighed by the value of having a completely free document editor with as many robust features as Google Docs provides.

Understanding the cloud-based nature of these products helps to explain some of the limited features, but the ability to access it from anywhere in the world on almost any device again, in my opinion, outweighs any shortcomings.

(When writing this book, I dictated a lot of the content to my phone into a Google Doc. Then I opened that file when I got to my computer and cut and pasted the auto-generated text into my manuscript software, which I figure saved me more than 100 hours of typing)

## GOOGLE SHEETS

Google Sheets replaces Microsoft Excel. Similar in look and feel just as Google Docs is to Microsoft Word, it retains all the same benefits that Google Docs has. You access it from the nine dots at the top right corner of your screen, or from the three dashes on the top left of the screen if you're already in Google Docs.

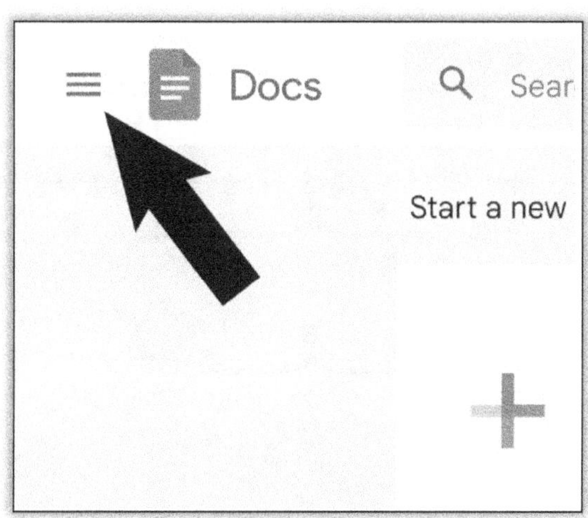

You see from those top three lines on the left that you can access Google Docs, Google Sheets, Google Slides, and Google Forms from the same screen, and they all share a similar look and feel when you open each.

As you can see, the layout of the user interface is very similar to Google Docs and has a similar feel to Microsoft Excel. As with Google Docs, the interface is familiar and functions very much the same.

Once you open the program by either opening a spreadsheet or a new file, you will notice the same similarities as Docs.

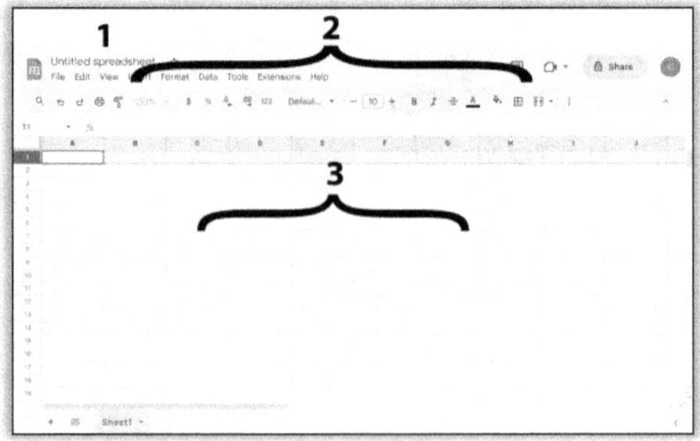

- 1. Your spreadsheet name and the File menu
- 2. Your feature/functions ribbon
- 3. Your blank worksheet

Google Sheets is not meant to be Microsoft Excel. It is meant to be Google Sheets. This means that you may find some of the specific features and functions you used in Excel you may not have access to in Sheets. However, if the features and functions of Sheets are sufficient for your needs, you just eliminated the need to purchase Microsoft Excel. This is a huge money saver for us as we don't use Excel for a whole lot, and having the free Google Sheets program meets all our spreadsheet needs.

Normally we are just taking a spreadsheet that needs to be converted to a .csv file and making the change in Sheets. Remember, you can access Sheets many ways, but if you just surf to https://sheets.google.com you will find Google Sheets.

## GOOGLE SLIDES:

Google Slides replaces Microsoft PowerPoint. This is a presentation software that you access by going to https://slides.google.com. You will notice it has the same look and feel as Docs and Sheets, but you are dealing with presentations now.

Once you decide to open an existing presentation or start a new one, you will be greeted with the familiar PowerPoint style interface, but remember, this is Google Slides! As we see in this image, we have:

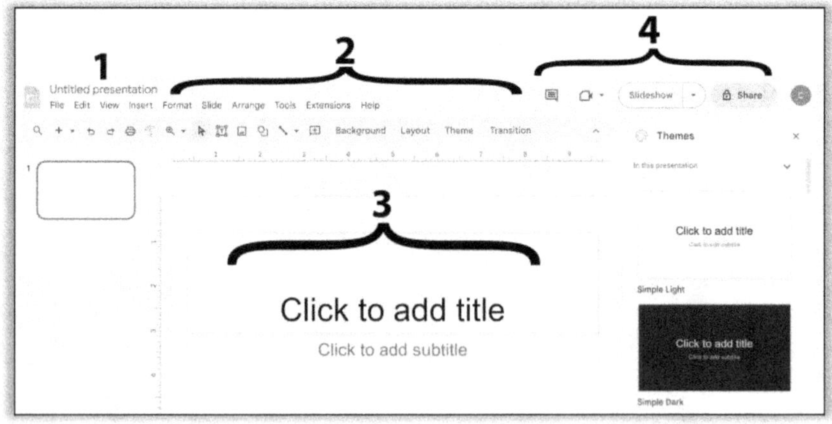


- 1. Your presentation name.
- 2. The File menu with the features/function ribbon below it

- 3. Your blank presentation, waiting for you to create your masterpiece!
- 4. This is your presentation area. You can see the version history, you can join a meeting and present this slide show, and you can share this presentation with others via the Share feature. Remember this file is cloud based, so if you give someone permission, they can view it without you having to send them a file, just a link!

This image also shows the templates (themes) that are available to you. There are a wide variety of themes to choose from, and you can customize any of them to get the exact look and feel you are seeking.

One of my favorite features of Google Slides is that I can open and view my presentation on any computer in almost any browser. I can show up at a meeting with nothing but a pen and my Google login information. If I am asked to present, I just login to my Google account and pull up my presentation. Then, to present I just need to click the "Slideshow" button at the top right.

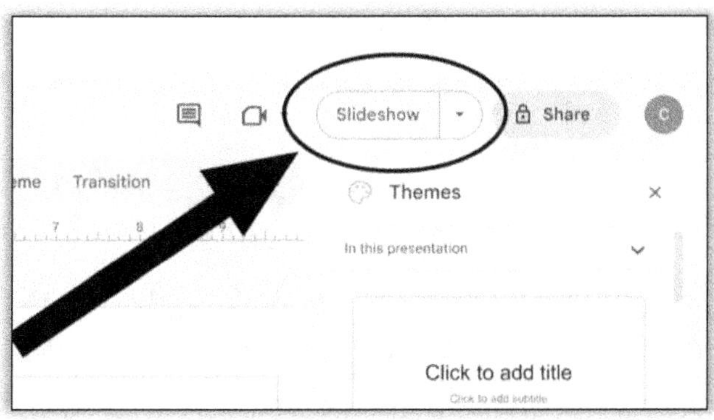

This is such a time-saving feature. I cannot tell you how many times I have tried to help someone who set their presentation up in PowerPoint and when they got to the location to present, there were so many different issues.

- Lost thumb drive.
- The computer they brought with them has a dead battery and no one else has PowerPoint on a laptop.
- The jump drive they brought isn't loading the presentation.
- This is the best. They forgot their laptop, and the person who offers theirs doesn't have the same fonts on their laptop and the presentation is a mess due to the replacement fonts not formatting the same.

The list goes on and on, but remember with Google Slides, if you have an internet connection, you are good to go. Clicking the "Slideshow" button will take your presentation to full screen and allow you to navigate through your slides like a pro. If you have a presentation "clicker" to advance your slides, this works with Google Slides as well.

So you can see the versatility of this program allows you the freedom to create a presentation on the fly and present it at the drop of a hat!

These 3 programs, Docs, Sheets and Slides can directly replace the Microsoft Suite of products you are currently using. I can honestly say that I used a 15-year-old copy of Microsoft Office (2003 version) for so many years because I couldn't justify the expense of purchasing Microsoft Office when I only used it a couple of times per year.

Now there are so many ways to purchase Microsoft Office that it completely confuses most people we work with. They have no idea how much they are paying or even what features they have or don't have.

When I found Google Docs, it changed my life! No more formatting documents and changing fonts. My documents and slide presentations just work! And they work well! This not only saves us the monetary cost of programs, but also the time and frustration savings of having all your documents at your disposal and no formatting issues when opening a 5-year-old file.

1. Do you have a local phone number?

_______________________________________________

_______________________________________________

2. When you are looking up a local company, if they have an out of area number how do you feel?

_______________________________________________

_______________________________________________

_______________________________________________

3. Would that cause you to call a different business?

_______________________________________________

_______________________________________________

_______________________________________________

_______________________________________________

_______________________________________________

4. What program do you use to write letters?

5. If you are using Microsoft Word, how much do you pay per year per user?

6. How many users are you paying for?

7. How much does that add up to each year? (do the math below)

8. What could you do if you had that much money still in your business account at the end of the year?

Remember, for updated information about this chapter, scan the QR code and watch the video for this chapter!!

George's story continued

"Rockstar," George started, "I keep getting calls from other junk removal companies. First they yell and complain at me because they say I am taking all their business. Then they say I am cheating the system and want to say I am doing something to mess with Google and some guy named Al. Then they usually say something mean about my mom….. and then they start asking how much it will cost them to buy my company!!!"

Well, what do you tell them when they ask how much you would sell for? "I tell them they can't afford to buy it. Then they usually cuss about how much they are paying for all the clicks that don't become customers. Then they hang up."

So, do you ever plan on selling? I mean it could mean a very nice retirement if the right company offers the right amount.

"Rockstar, if I am ever going to sell this company, it will be to fund a homeless outreach in our local community. I don't need a lot, and any extra I get will be used to Glorify God."

Of course it will be George. Of course it will.

# MAXIMIZING GOOGLE PHOTOS

"A picture is worth a thousand words"
— Fred R. Barnard

My next favorite feature of Google is the Photos feature. Google Photos has proven to be a huge time saver when dealing with photos.

I used Evernote in the past for all my estimates and designs with my sign company. I could take photos with my phone in the Evernote app and then I could log into Evernote on my computer and have access to all the photos.

However, if I didn't remember to take the pics *in* the Evernote app, I would first have to go to Evernote on my phone and add the photos to Evernote. Then I could login to Evernote.com on my computer and find my pictures there.

This all changed with Google Photos. As an android phone user, I have my phone automatically backup all my photos to my Google Photos cloud storage. Every picture I take is immediately uploaded to Google Photos.

This convenience has saved me countless hours looking for a picture or remembering where I was saving those pics. Not only are they saved forever, but Google has some incredible search features within the Photos app and website so you can look up your photos from 10 years ago by searching the location where the pic was taken or something or someone who was in the picture. Really. Try it. Go to https://photos.google.com and login to your Google account.
There you will see every photo you have ever saved to your Google Photos. At the very top is a search bar, type in any term (pizza for instance) and look at the results. Pretty impressive, huh? Every single picture in the results has pizza in it.

Google Photos is an amazing free storage area where they give you 15 GB of storage at absolutely no cost. This 15 GB does include your Gmail and all those attachments as well as everything else in your Google Account. When used properly, there's almost no limit to the number of photos you can store in 15 GB.

You can always check on how much storage you have and whether it's being used by your photos or your Google drive and Gmail by going to https://photos.google.com and clicking on the small gear at the top right corner.

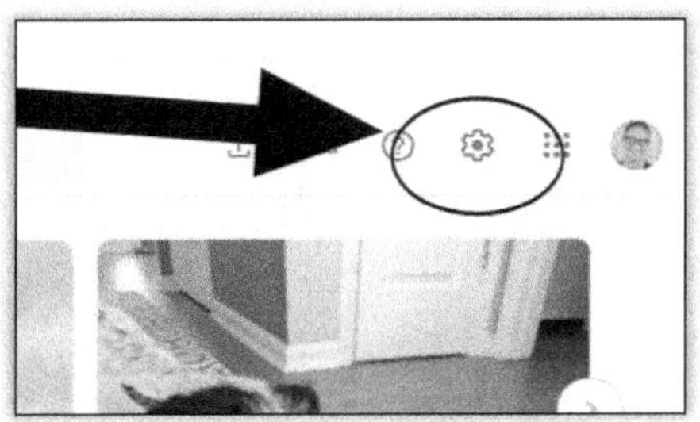

This will allow you to see what quality you have selected to back up your photos and videos. If you click the

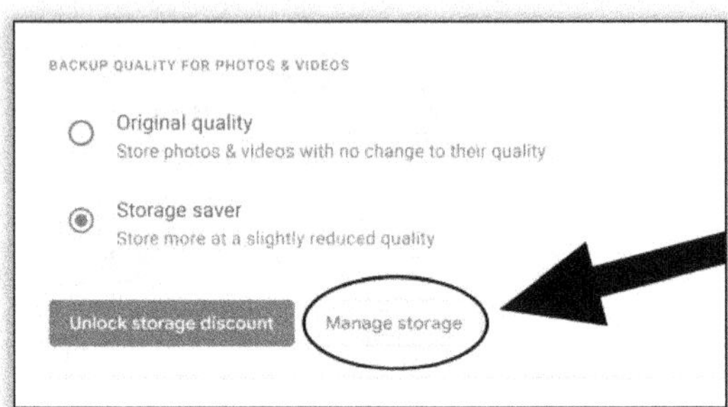


"Manage Storage" button, you will see a page showing you exactly how much of your storage is being used by each of the above-mentioned items.

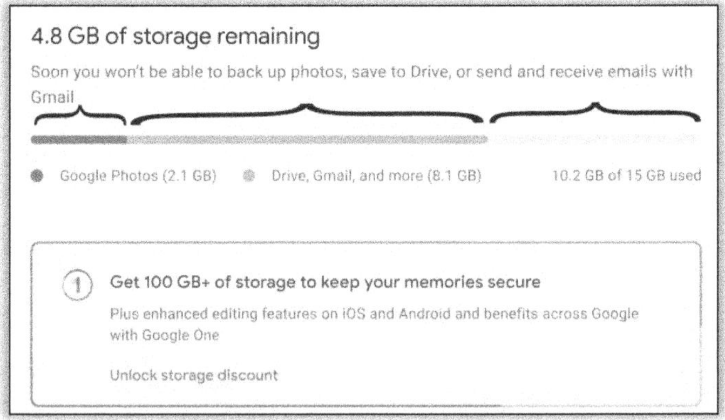


From the default photos screen, you can see every photo that you have taken. The more you scroll, the farther back in time you'll go. If you hover over the scroll bar, it becomes a date range, and you can see the dates getting farther back as you scroll. As you continue to roll your mouse wheel, you'll see the dates jump farther and farther back. If you click and hold, you'll go very rapidly through the date range.

Once you've clicked on a photo, you'll see a menu at the top right corner. Click the i in the circle. This is the info icon.

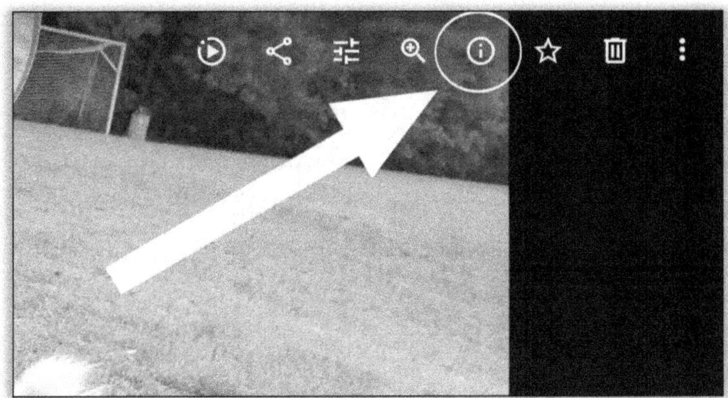

The info icon will allow you to locate all the information about the photo, including the date, time location. (if the device that took the photo had location turned on). This is helpful when searching for specific event photos if you can

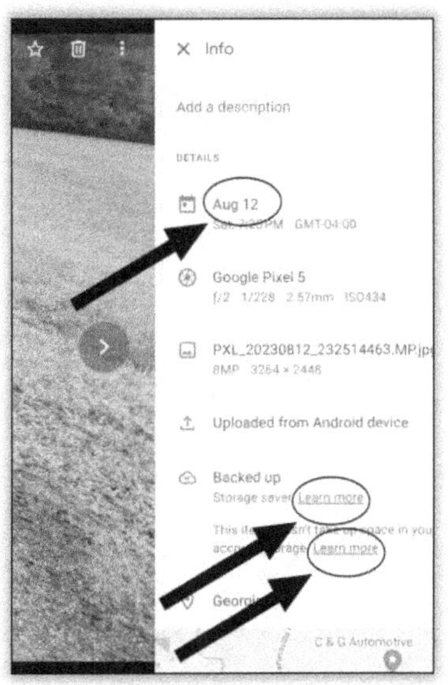

remember where they were taken, or an estimated date to look for.

Also, while on that screen you can click on some of the learn more hyperlinks to learn about Google photos backup and what counts towards your storage and what does not. Google is good at keeping the contextual links for the information right where you need it and having the information right there to click on and learn about your photo quality and how much space is being taken up is a good placement of that information.

Not having to pay for this storage area and being able to access the photos from the desktop instantly upon taking them provides quite a bit of value to our company. We do a lot of our own social media posting as well as advertising flyers and if we want to include a photo in it this makes it extremely easy.

If you need to download your photos, navigate to the photo you want, and click on the three dots in the top right corner, and there is a download option on the menu.

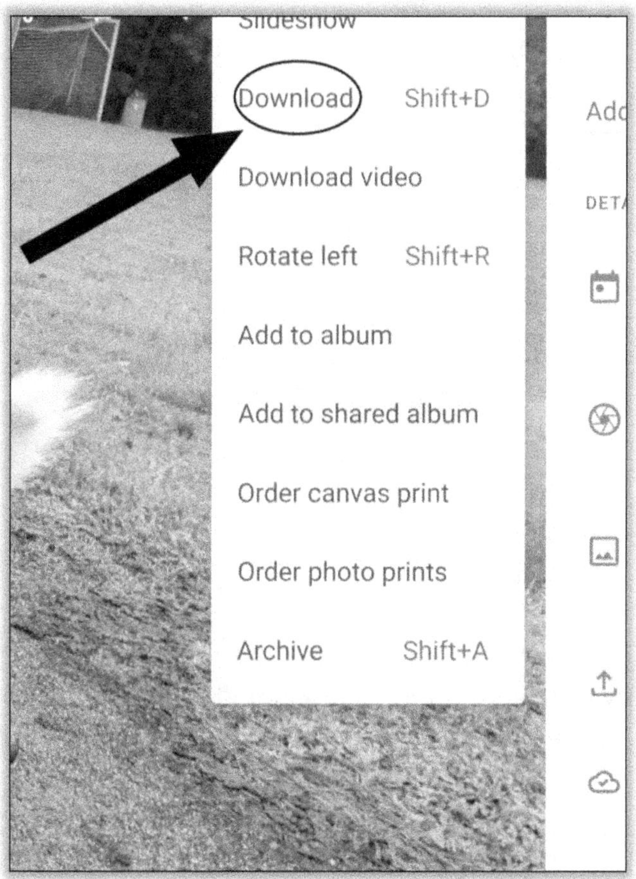

Google Photos works with lots of third-party platforms. With a lot of them, you can connect to your Google Photos directly from their platform. One of these is Canva.
They have a wonderful website for all things graphics related. You can log into Canva with your Google account and then access your Google Drive and Google Photos directly from inside your Canva designs.

This is just another example of third-party companies wanting to be able to use Google's size to gain access to their customer base. Allowing customers to log in using their Google account and then allowing their Google photos and Google drive to be seamlessly integrated into that platform is a terrific way to create a relationship with Google and to make your customer's journey very simple. This lends itself to large customer sign up and retention as people want ease of access and ease of use. Sorry Microsoft users, there is no Canva connection for your cloud storage options.

There are so many features available with Google Photos, it would take another book to talk about them all. From editing your photos to making them indexable, creating slideshows and ordering prints, Google has countless options for you to access from within the Google Photos product.

1. Apple or Android?

2. If you are on Apple, do you have the Google Photos app installed?

3. Think of the last time you wanted to get a photo off your phone. How did you accomplish that?

4. Have you ever used Canva.com? List 3 ways you could use some of the pics off your phone to create some cool card, flyer, or social media post.

Remember, for updated information about this chapter, scan the QR code and watch the video for this chapter!!

George's story continued

Teaching is a passion for me. However, when someone already has their own idea of how to do what you are trying to teach them, this can prove to be frustrating.

I have many customers who come to me for my training. They pay me a fair price to walk them through setting up their Google Business Profile and then continue to do it "their way" as they go forward. They don't take the time to do the "homework" and they certainly don't think anyone knows better then them how to market their company.

I bet some of them walk away from my training saying they already knew most of what I taught.

And then there is George. He didn't have any other option. He trusted me, my system and God, and he did what I showed him to do.

And he built relationships. Lots of them. He followed up with previous clients and would randomly text customers to check on them. This was marketing genius (wonder where he got that from) and helped to build his business on a firm foundation.

# LEVERAGING YOUTUBE (OWNED BY GOOGLE)

> "Give me a lever long enough and a fulcrum on which to place it, and I shall move the world."
>
> — Archimedes

YouTube is the single coolest completely free product Google offers. There. I said it. YouTube is so cool that I literally jump around when I am presenting live about Google and YouTube. I just want business owners to grasp the power of what Google has given us access to.

Not only do you have a place where you can post videos of almost any type of content, but you have no financial or custodial responsibility to maintain the hardware that those videos are being stored on. You aren't responsible to pay for the internet connection for those videos to be accessible to the world, or to update the software that keeps those computers running where that video is hosted, or the electricity required to keep those computers running!

Google allows you to post unlimited video content on your private YouTube channel. This can be customized to your business and then used to promote your business through the fastest growing content on the planet. Video is never going away and to be able to deliver your company's message in a concise way is paramount. Having a platform where you can post your content and then share it with others or embed it in your website is an invaluable service for a small business.

I can't explain emphatically enough how difficult it was to try and put a video on a website in the past. Nowadays you just go to your YouTube channel, find the video you want to put on your website and then copy the share link from that video. Then you post that on your website's YouTube field and voila! You have embedded video playing on your website.

We use YouTube for many things, Building a brand image is the most important part of the YouTube piece of Google's Free S#!t. Having short videos that clearly state your message to the world is helpful, but offering short videos that give answers to your client's and potential clients is an even better strategic move. As you build your brand image by getting people to come and look at your explainer videos, more and more people will come to see you as the expert in your field. Then, when they do need help with something that you are talking about on your video channel all the time, they're going to remember that you were so helpful in the past and then they will search you out to use you as a paid solution to their problem.

Once you go to https://youtube.com, you will sign in with the same Google account that you've used all through this book. Once you've signed in, you'll be greeted with a screen showing a bunch of suggested videos with a search bar at the top and a menu down the left-hand side.

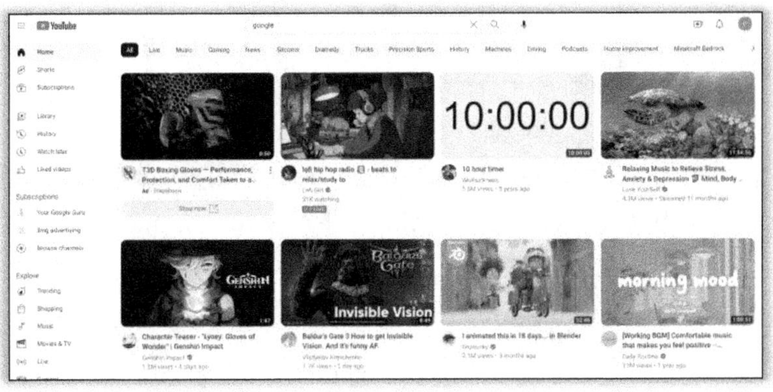


You'll notice your Google account logo in the top right corner next to a notification bell and a projector with a plus sign on it. You'll access all your YouTube account information from your Google account logo at the top where you can see your channel and go to YouTube studio and switch to other YouTube accounts if needed.

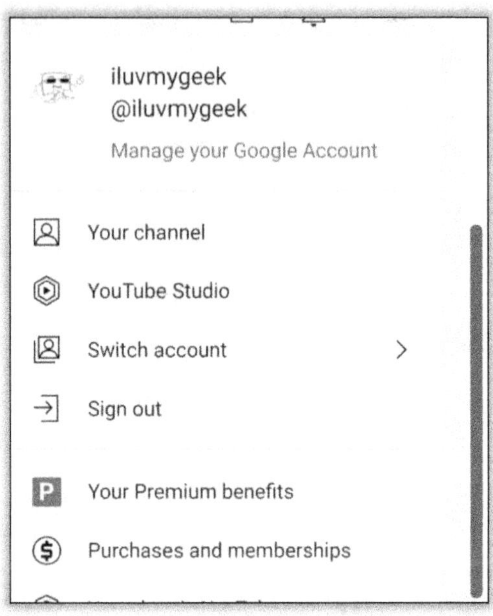


We have multiple YouTube accounts for all sorts of different content that we want to display. Also, you can access all your settings for your YouTube account from there. If you want to see all your videos you would click on your videos / content on the left-hand side of the YouTube screen, which will take you to your channel content page showing you some overview information about the videos that you have on your YouTube channel.

Not only can you use YouTube for publicly posted videos, but you can also use it for internal training videos and videos that you're sharing amongst your employees. We make training videos for our Google customers and then put them on

YouTube so the client can access them at will when they need help and some assistance remembering how to do something that we show them in the training.

The following pages assume you have a YouTube account already. If you don't and you would like some help setting one up, contact me.

To upload a video to YouTube, you need to log into your YouTube account. In the top right-hand corner where you see your Google account information, you'll notice a notification bell and a small projector with a plus sign.

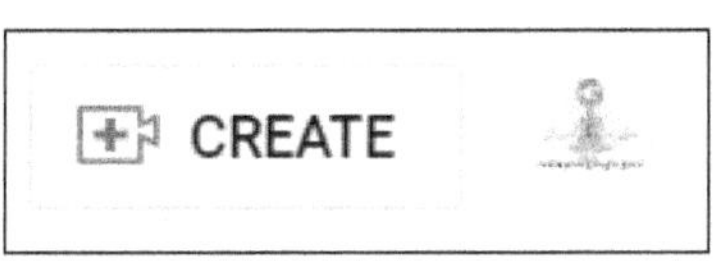


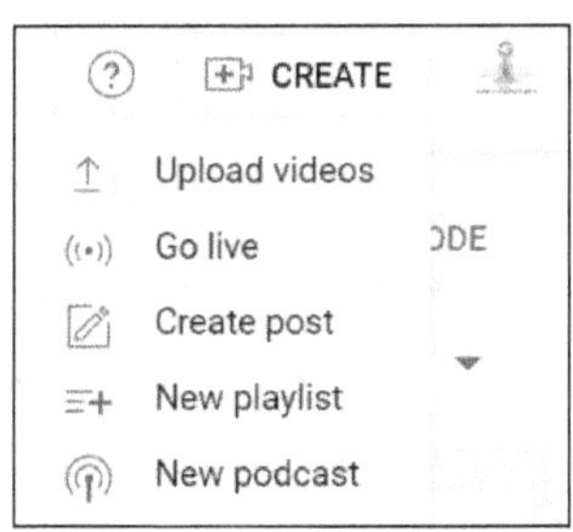


Click the projector with the plus sign and your options are to Upload video, Go live, or Create post. This is how you engage with your audience through YouTube.

We're going to go over uploading a video so click on the Upload videos menu item and you're going to be presented with a window that says drag and drop your video or click to select your files.

Once you've selected the video that you would like to post on YouTube, it will start uploading and you will be presented with the details screen. This is where you get to tell the world about your video. Your title gets to be up to 100 characters, and it should be very descriptive. Then you have a description area where you can go crazy with up to 5,000 characters about your video.

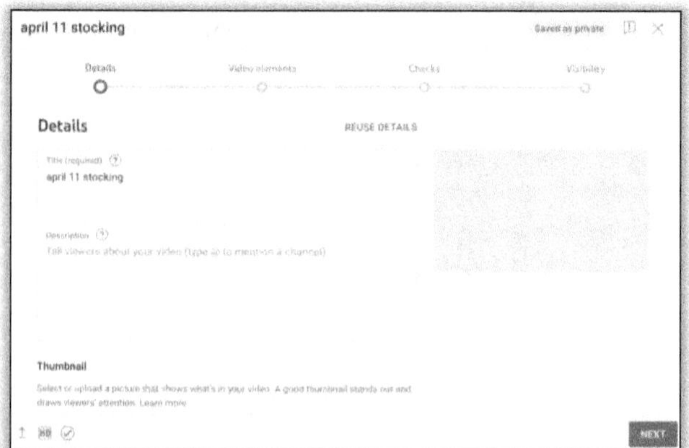

We don't suggest you use 5,000 characters; however this is where you would put in any of your hashtags and any of your links that you wanted to share with your video.

As you scroll down, you will be able to choose a thumbnail. Thumbnails are an important part of all YouTube videos as they are how you get people to click on your video when they're scrolling through YouTube. If you do not create a compelling and inspiring thumbnail you will not get people to click on your video. YouTube will take a few of the clips from your video to create a thumbnail and offer you three options, but it is a much better idea to build yourself a really nice thumbnail in Canva and then upload that when you're on the details page of your video.

You can also add your video to a playlist if you would like. Playlists are helpful when you want to show multiple videos in a loop on a screen somewhere. For instance, if you had a TV running commercials in a local restaurant and you wanted to be able to change out those commercials regularly you could use a YouTube channel playlist. Then you would go to the device that was going to display the videos and go to YouTube and loop that playlist continuously. Then from anywhere in the world you could add videos to that playlist and remove videos from that playlist with no further interaction of the people on site at the restaurant. The playlist would update automatically on their end and the new videos would show and the ones that were removed would be gone.

The audience should always be set to not made for kids unless it is a child's video. Then, you have some other restrictions and compliance that you need to make sure you take care of. All of my content is made for everyone, but not specifically for kids, but we say no it is not made for kids and that removes any of the additional compliance features.

There is a show more button that will allow you to put in subtitles and tags, however for this book we're not going to go that deep into YouTube.

When you click "Next" in the bottom right-hand corner, you will be presented with the video elements screen, which will

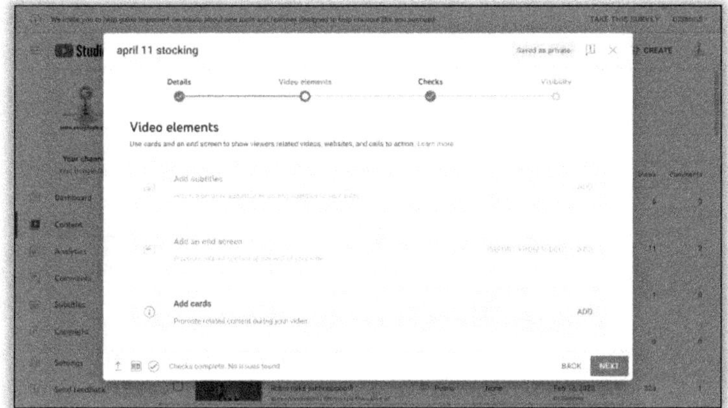


allow you to add subtitles and an end screen or cards to your video. Each of those has a description on the screen.

Clicking "Next" brings you to the checks screen which is where YouTube will let you know if it found any copyright problems or digital rights problems with the music that's in your video.

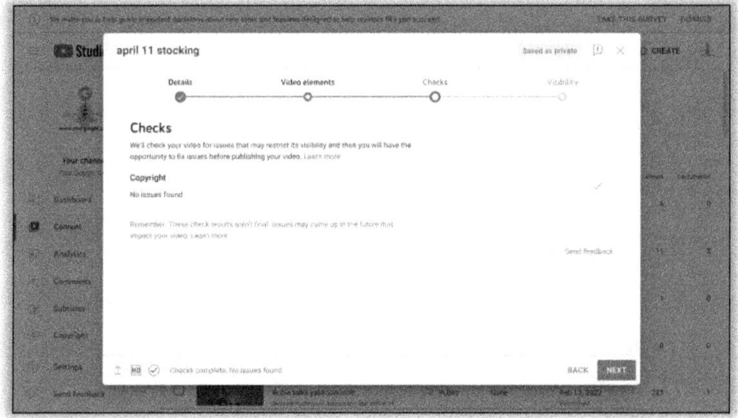

Hitting "Next" one more time will bring you to the visibility screen where you get to choose if your video is going to be private, unlisted or public.

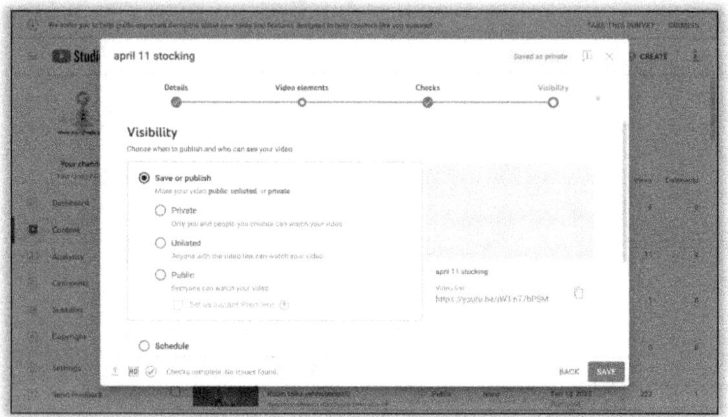


Once you save the video there, you will be presented back to your channel content screen and see your new video on the list of videos.

I am so amazed at the power that this free resource from Google holds, and I could go on and on about the different parts and how much each is worth. Unfortunately, that could be an entire book on YouTube and we're not here to discuss that. We're here to help you understand that you have the ability to post as much digital video content as you desire on a public or private platform. This does have some possibilities for financial gain if you produce great content.

After getting a certain number of video view hours per year, YouTube will start running commercials for other companies on your videos. Once you reach what they call the "monetization stage", you get to share in the profit from those commercials. There are some pretty stringent requirements, but once you have gotten past those, you can make a very nice income creating great video content and posting it on a completely free platform called YouTube.

I am not here to discuss how lucrative YouTube can be, but go ahead and Google "Blippi the Clown net worth" and you will get some idea of the possibilities!

1. Have you ever wanted to use video to explain how to do something?

2. Have you ever used YouTube to find out how to do something? What was it?

3. List 3 ways you could use video to gain more customers. What type of video would you create?

4. List 3 inspiring titles for those videos

Remember, for updated information about this chapter, scan the QR code and watch the video for this chapter!!

George's Story Continued

As the years went on, George started using Twitter. He found many commercial clients by commenting on people's tweets and he created a small following. I can remember sitting in a meeting with George and he was texting away under the table.

"What are you doing George?"

"Giving an estimate for a job on Twitter" George replied.

"Twitter? How did you hear about Twitter?"

"Well Rockstar, my son said that lots of people use Twitter, and the President likes to Tweet, so I decided maybe I could get some business on there and I started a Twitter account."

You could have knocked me over with a feather. My friend, my very non techy friend, was using another online platform to promote his business. I have never been so proud of someone that I wasn't a parent to.
Great job George …

# MEETING GOOGLE MEET

"Being in the same room with people and creating something together is a good thing"

—Robin Williams, Actor

In the spring of 2020, the world went crazy and we all had to go home and do all of our work virtually. Video conferencing became mainstream. Kids needed it to attend school. Mom's needed it to have appointments and talk to other moms. Workers needed it to stay connected to work and most importantly, businesses needed it to continue doing business.

When all this stuff went down, a small video conferencing company that had been around for years came to the forefront and suddenly Zoom calls were an everyday event. School was over Zoom, work was by Zoom, networking meetings went to Zoom and the "virtual meeting" became the norm.

Of course, this was a huge financial boost for Zoom, and they have taken their software platform to great new heights. They are now probably the leading video conferencing company in the world, and the latest figures say they are hosting approximately 55.8 Billion hours of meetings per month.

Pretty impressive, huh? There is no doubt that they have a solid and robust platform. I used Zoom from March 2020 until the middle of 2024. I was hosting a weekly meeting with a group called Happy Neighborhood Project, and they had a deal for the hosts. If you joined their team, you were able to purchase the $250.00 monthly Zoom package for only $12.99 per month! This was a great resource for my business as I do quite a bit of online meetings and trainings, and I like to have a consistent link to share as well as I really need the program to automatically record all my meetings. I will always forget to start the recording if it doesn't start automatically. Making a recording of each training session to share the video with my client so they have something to refer back to is critical, so having it record for me automatically is a requirement.

Well, in the middle of 2024 we were informed that the deal with Zoom was going away, and we would have to pay the actual full price of $250 per month. Bye bye Zoom…..

Hello Google Meet!

Google Meet is the Google version of Zoom. Well, not totally, but a pretty nice video conferencing platform that you can use as part of your Google account, of course, for FREE. Yes. As with almost everything I am sharing with you in this book, Google Meet has a free version, and if you decide to use any paid offerings from Google ( like their 2Tb of storage for $99 per year) you get access to a bunch of upgrades such as longer meetings and recording options.

As with all the other cool free s#!t from Google, Meet is accessible from the Application Launcher (aka "the waffle" at the top right of your browser window.

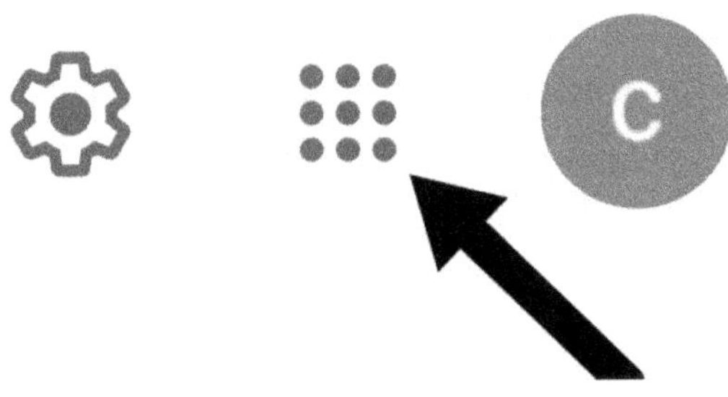


And then select the meet icon:

This will open the Google Meet page so you can access Google Meet. One of the great parts of using this platform is there is no download and installation. All you have to do is sign in and you can start using Google Meet

You simply click the New Meeting button, and you will be offered the option to:

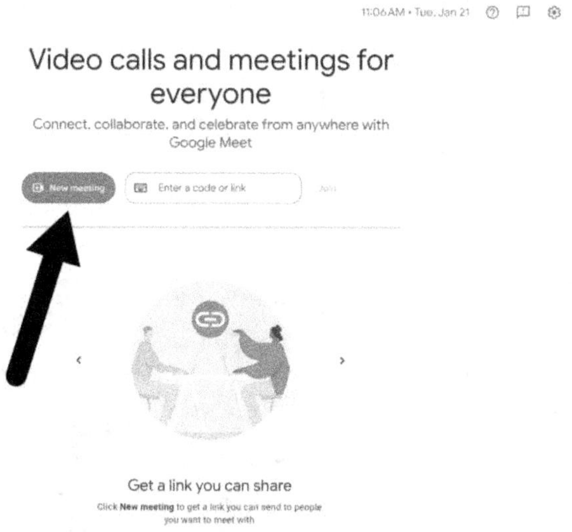


Create a meeting for later

1. Start an instant meeting
2. Schedule in Google Calendar

Let's go over the 3 choices. The first is Create a meeting for later. This will just give you a link that you can send to whoever needs to attend your meeting. This is a simple way to just get what you need and get going. The second choice is to Start an instant meeting. This will immediately start a virtual meeting after asking you about the camera and microphone. Schedule in Google Calendar is the third choice, and we will go over some parts of that towards the end of the chapter. Once you click Use microphone and camera, the meeting will begin

When the meeting starts, you will be prompted with your meeting info and the option to Add others or copy the meeting link. Google has made this as easy as possible, and it doesn't need to be hard. Just copy the link and send it to anyone you need in the meeting. You can text it or email it. If you have an email address for the people you are inviting, you can click add others and put in that email address. It will send them a notification and or email to let them know it is time to join.

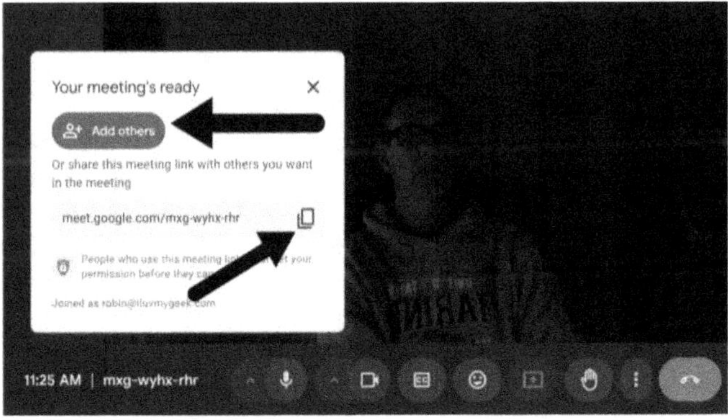


Now that you are in the call, let's go over the icons at the bottom of the screen. From the left to right you have your time of day, then you have your meeting code. I will talk

more about this later. Next item as we move to the right is the microphone control. This is so you can choose and control your mic and speakers. (some of us have multiple mics and speakers) and you can also access the settings for Google Meet from this menu.

As we move to the right on our main control bar, the next item is the video control.

The video controls are very much like the audio. Same option to choose your video source and the gear icon to access Meet settings. There is also the portrait image icon that is for accessing the "apply visual effects" menu. We will discuss that a little more later.

Next on our bottom control bar is the cc or closed captioning control. This is just what it says. It provides real time captioning for viewers and can be used in a variety of languages. Once you turn it on by clicking on the cc icon, you can click the language (displayed in blue) and choose your language. It won't translate, so if you are speaking in English and want it to caption in Mandarin Chinese, it won't work. It is a captioning service, not a translator.

Next is the reaction control. You can toggle your emoji bar on and off by clicking on the smiley face control on the bottom bar.

Then we have the Present Now control. This one is a little more involved, but worth the time to go over it so you can be fluent in using Google Meet.

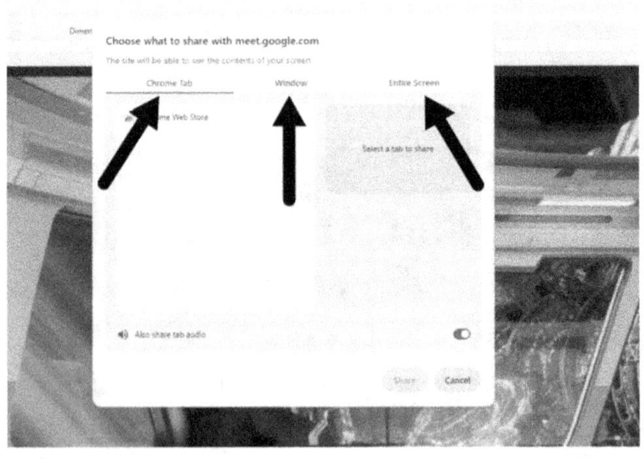


There are 3 options when presenting on Google Meet. The first option is "Chrome Tab". This will allow you to show a specific tab on your Google Chrome browser. This is convenient if you are only going to show one tab to your audience as it still allows you to navigate to other tabs while the audience only sees the one you are presenting.

This can be frustrating if you change tabs while you are presenting to an audience and you forget to share the active tab. Now you are talking about the tab you see, but the audience is still looking at the tab you shared. In that case you would want to choose the window or the entire screen option when you start your present now session

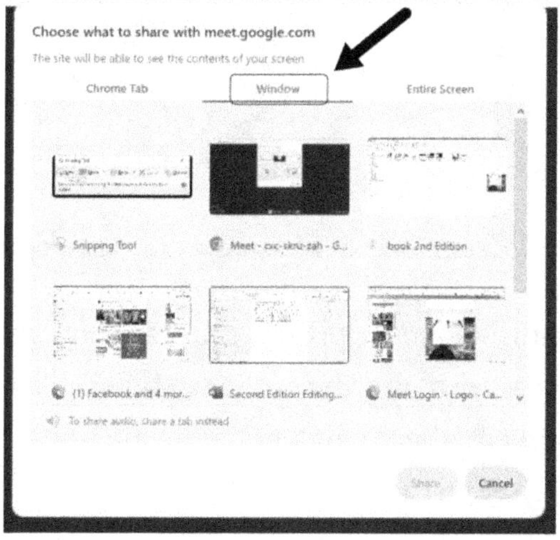


With the window option, Google Meet will display whatever is in the window you choose. That could be any program you have open, from Chrome to Word to snipping tool to whatever you want. If you show your Chrome window, you won't have to remember to display the new tab when you change tabs, Meet will show whatever is in the window, so it doesn't matter which tab is active, what you see is what your audience sees. If you are working on editing a Word document, you could show your Microsoft Word program (window) and your viewers would only see what you were doing on that program.

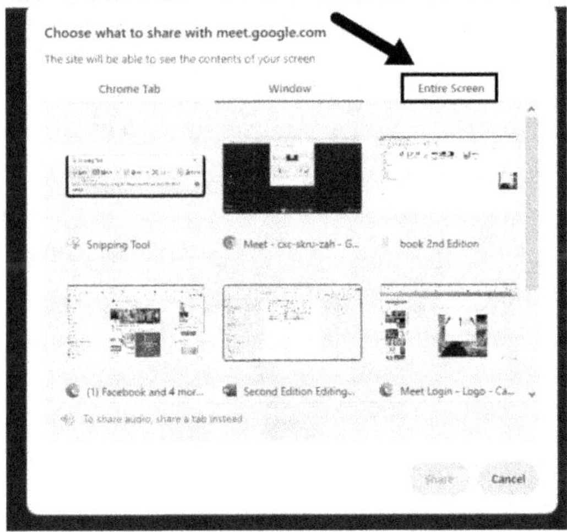


With the entire screen option, you can put whatever you want on your screen and the audience will see it. If you only have one screen, then whatever you are looking at, your audience will see as well. If you have multiple screens, you can show what is on one screen while viewing your audience on another. It is very similar to other video conferencing platforms.

The next option on the bottom bar is the raise hand option. This is self-explanatory; it will show the meeting you have raised your hand. You can always just start a Google Meet meeting by yourself and try out all the features. There is no charge for using this program. Just be signed into your Google account and head to https://meet.google.com.

The next item on the bottom menu is the 3-dot stack. This will bring up another menu to access items such as changing the layout of the people on your screen, going to full screen or popping the meeting out to a smaller window with picture in picture. You can also access the apply visual effects menu as well as the legal and technical stuff that helps us keep this all working.
Finally at the right side of the screen at the bottom you have a smaller menu with the Meeting Details, People, Chat with Everyone, Activities and Host Controls. These are all self-explanatory and if you have any more questions, head over to our Google page and ask!

Ok. I said I would come back to the third option when you start the meeting and that is Schedule with Google Calendar. This is probably my favorite option, and it makes life very cool when you are using virtual meetings for 90% of your business.
I made a big deal out of being able to set the meeting to record automatically. This is only possible if you schedule a meeting with Google Calendar. I also said I like to have a constant link that never changes. This is also only available when scheduling a meeting with Google Calendar. So here goes.... If I schedule a meeting with Google Meet, it is going to make a new meeting code every time. This won't allow me to have a cool, custom link for my clients to use to get to my meeting.

So here is how to make this work. I need a "permanent" link for my meeting, however, every time I schedule a new meeting, Google Meet issues a new meeting code. But if I schedule a meeting for 10 years from now I will get a meeting code that is good for the next 10 years!

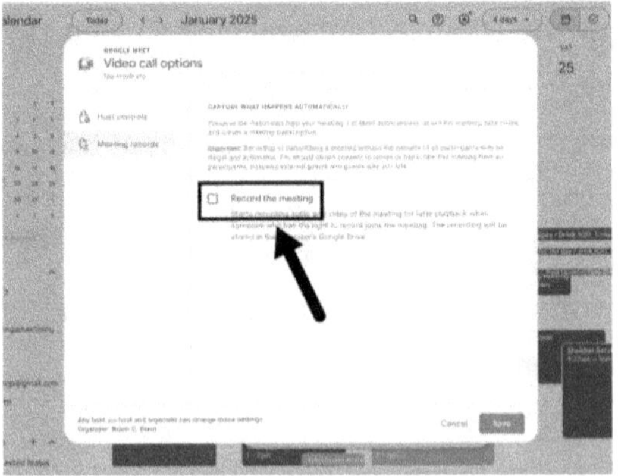

Annnnnd, when I am scheduling that meeting, what if I check the box to "record the meeting automatically" so I don't have to ever remember to click the record button? Problem solved. I now have a "permanent" link, and all my meetings will record automatically. Then I can use a redirect to have a cool URL for my meetings that I point at that Google Meet link.

This has revolutionized my coaching business, and it is one more free thing Google has given us to use to be successful in our entrepreneurial journey! If you want to try it out, go to https://meet.robincstern.com

1. Have you ever been to a virtual meeting?

___

___

2. If you have been to virtual meetings, which program or platform did you have to use?

___

___

___

3. If you had to run virtual meetings, what features are important to you. List 3 things you would like your meeting platform to be able to do.

---

---

---

---

4. List 3 things that you would like to share with the world that you could use virtual meetings to accomplish.

---

---

---

---

Remember, for updated information about this chapter, scan the QR code and watch the video for this chapter!!

# GETTING IT ON WITH GEMINI

"I'm sorry, Dave. I'm, afraid I can't do that."

—Hal 9000 from the movie *2001: A Space Odyssey*

Google Gemini is our next free thing from Google.
As with all the other cool free s#!t from Google, Gemini is accessible from the Application Launcher (aka "the waffle" at the top right of your browser window.

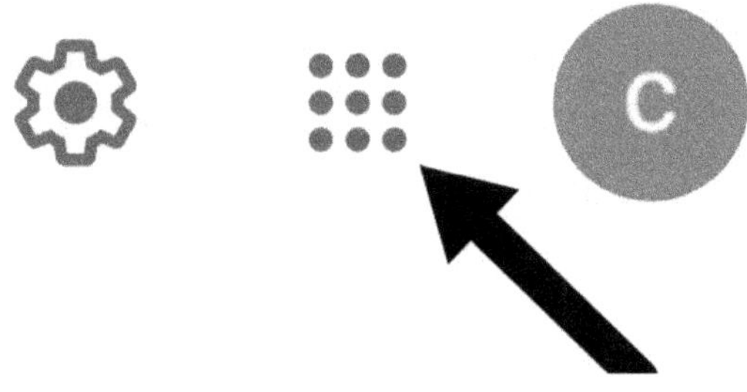


And then select the Gemini icon:

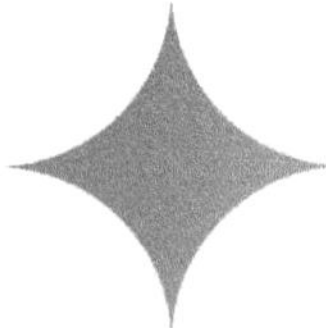

Gemini is the Large Language Model that Google has released that allows us to interact with AI on the Google platform.

Let's ask Gemini what it is. The following paragraph was generated by Google Gemini from the following prompt:

"Explain to me what Google Gemini is" and here is its response:

*Google Gemini is a multimodal AI model developed by Google. It's designed to understand and combine different types of information, including text, images, audio, video, and code. Gemini comes in different sizes (Nano, Pro, and Ultra) to cater to various user needs, from quick on-device tasks to complex, large-scale operations. It has advanced coding capabilities, including code translation, problem-solving, and code completion/fixing. Google is integrating Gemini into its products, such as Pixel 8 Pro and Google Bard. It demonstrates exceptional performance across a diverse range of tasks surpassing human experts.*

So now you are saying, ok, what the heck does that mean?

I felt the same way. So, I prompted Gemini with this request:

"Now say that so a fifth grader could understand"

And this was the result:

*Okay, imagine Google Gemini is like a super-smart robot helper. It's not just good at one thing, like reading or drawing. It's good at EVERYTHING! It can understand words you type, pictures you show it, sounds you make, and even videos.*

*Think of it like this: you have a friend who's amazing at LEGOs. They can build anything you describe! Gemini is kind of like that friend, but instead of just LEGOs, it can work with all sorts of information. It can write stories, answer questions, and even help you fix problems with computer code!*

*There are different sizes of Gemini, like small, medium, and large, depending on the problem, and how much help you need. The small one can fit on your phone, and the really big version is super powerful. Google is putting Gemini into lots of its products to make them even better!*

Wow. What is that all about? Well, just think about it as an assistant that can search millions of sources in a matter of seconds. Data, images and a myriad of other information can be accessed in seconds by your "personal assistant".

This can be used for all sorts of different results for your business. For instance, we use it to generate social media content and give us information about our perfect customer. We like to ask it to search the local area and tell us who is the target customer be for a local computer store that sells refurbished products. And I asked it for what sort of arguments those potential clients would have about using a local store over a big box store, and purchasing refurbished technology. Not only did they give us some information about our potential clients, but it gave us all the arguments that they would have about using our business for purchasing or getting a computer repaired. Then we ask it to give us answers to all the arguments those pension clients would have, and then it takes all of that information and creates social media posts for us including hashtags emojis and punctuation.

Let's see how we access Google Gemini. Go to https://gemini.google.com, and if you're logged into your Google account you will be able to immediately start prompting Gemini. You'll see a screen that looks like this:

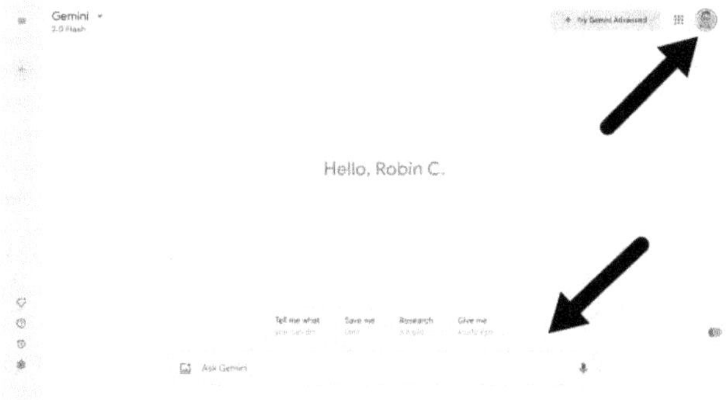


Notice the top arrow pointing at my currently logged in Google account and the bottom arrow pointing at the prompt field where it says, “Ask Gemini”. This is where you get to input your question / prompt / request. Gemini will also allow you to upload images and speak to it using your microphone.

This has been a tough chapter to write due to the constantly changing face of AI in 2025. Gemini is the name of the AI model; however, Google has found so many ways to incorporate it into a multitude of products and services. Pixel phones have Gemini built in. Google Apps (Docs, Sheets and Slides) are offering Gemini as a constant assistant to help with content, formatting and research.

Everyday there is a new feature or platform where Google has offered us a new way to interact with Gemini. Just last week I was exposed to NotebookLM. This is one of the coolest interfaces yet. You can upload "sources" of course Google Docs, Slides and Sheets are available, pdf, text and mp3 files as well. You can also give it links to websites and YouTube videos or just copy/paste your text right in. Then you can direct the output with steering comments and any customizing you like. But the absolute best part is you can generate a "Deep Dive" conversation that is podcast quality. Yes. Gemini can generate a conversation between a male and female voice that will "discuss" your sources and give you an audio file output to use for any purpose you need.

You can also ask for a Study Guide, Briefing Doc, FAQ or Timeline of the discussion to create any type of output you need. This is revolutionizing our online world and allowing us to interact with Google Gemini in a whole different way. There are so many different possibilities, truly your imagination is the limit. Let's just say I will be updating the video for this chapter on a regular basis to keep up with

Google and their creative innovations to allow us to use AI in new and intriguing ways.

1. Have you ever used AI? If so, which company's product did you use? (ChatGPT, Perplexity, CoPilot, Gemini)

________________________________________

________________________________________

2. List 3 things you would like to use or have used AI to accomplish.

________________________________________

________________________________________

________________________________________

3. Do you have an AI assistant in your home? What sort of things do you ask it to do?

________________________________________

________________________________________

________________________________________

________________________________________

4. List 3 TV shows or movies you remember that used some sort of AI or robotic interface to interact with humans.

______________________________________________

______________________________________________

______________________________________________

______________________________________________

______________________________________________

______________________________________________

Remember, for updated information about this chapter, scan the QR code and watch the video for this chapter!!

Afternote about Google Gemini and AI in general.

To really keep up with what is happening in the world of AI, one would have to spend most of their days researching and trying out different AI interfaces. This book is not meant to be the final word on something that is constantly changing. However, we would love to invite you to connect with us on any of the social platforms and follow along.

I talk about AI a lot on LinkedIn, but you can follow along on Facebook or YouTube as well. There is so much to come, and it is changing so rapidly that I would be remiss if I didn't give you some way to keep up to date.

As a simple way to find me and all the resources I have, just Google #robincstern or #yourgoogleguru and you will be able to access all my socials as well as any other resources I have out there. (There is some bonus material in the back of this book that will explain the hashtag thing if you are interested)

# USING THE REST OF THE FREE S#!T

"Give It Away, Give It Away, Give It Away Now..."

—Red Hot Chili Peppers

## GOOGLE CHROME / GOOGLE SEARCH / GOOGLE MAPS GOOGLE ASSISTANT

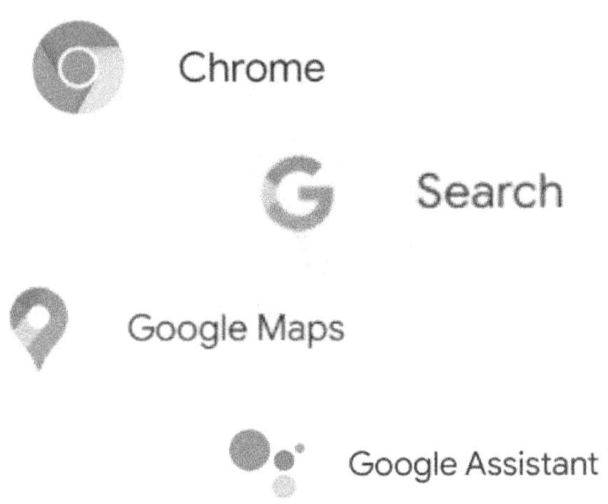


When discussing all of the Free S#!t from Google, we would be remiss if we didn't mention Google Chrome, Google Maps, Google Search and Google Assistant. These are the four that are part of our everyday lives. Most of us use these 4 programs so much they have become invisible. We barely know that we're using them. I'm using Google Assistant right now to write this book, just hitting the microphone button on my phone and speaking to it. Google Maps is constantly in the background helping us with GPS to find our locations and our destinations, and Google Chrome is the most widely used web browser in the world. Check that information. Even if you use Apple products, the most frequently used search engine is Google Search. These items are so much a part of our lives that we don't even notice them, so I wanted to make sure that I gave them a mention before we wrap this thing up.

## GOOGLE CHROME

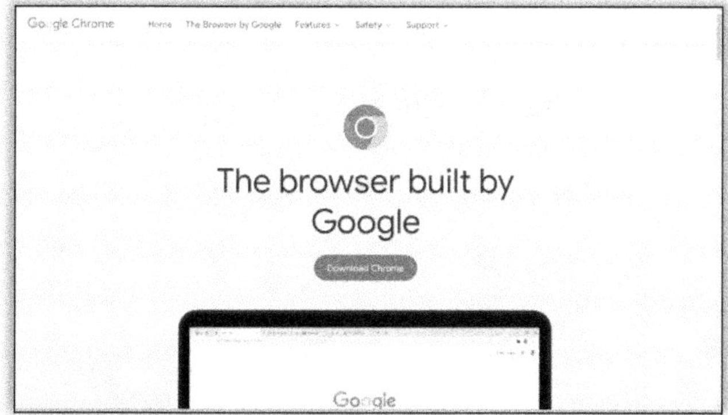


Google Chrome is the web browser program from Google. This program allows you to access the internet and browse webpages from your computer, tablet, or phone. Google has integrated many cool features into Chrome, and we take so many of them for granted. Once you log into your Google account inside of Google Chrome, you have access to all your Google products from the dot array (the "Waffle" in Google speak) in the top right corner:

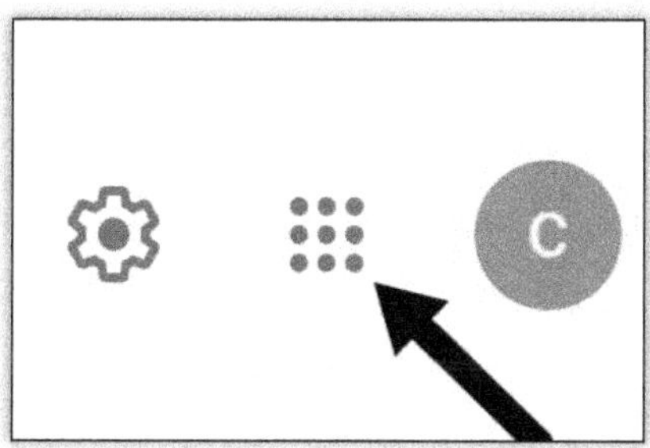


Here you will see many Google offerings, with the most frequently used programs at the top of the list. Gmail, Calendar, YouTube etc. are all accessed from the "Waffle"

and since you are already logged into your Google account, all your documents, slides and spreadsheets are now available to you in each program. Also, Google Drive is located here, and you will see that you are already logged in there as well.

Chrome has many useful accessibility features. The browser supports screen readers and magnifiers and offers full-page zoom, high-contrast color, and text to speech extensions for people with low vision.

Google works very diligently to keep updating and innovating all their products, and Chrome is no exception. As one of the most used Google products, they make sure the user experience is top notch and always cutting edge. Google's Chrome browser is also among the few products used by a billion people worldwide. Available on desktop as well as mobile devices, it is the browser of choice for many.

Across all operating systems, devices and platforms, it has a market share of 65%.
Next, let's look at one website you could visit from your Chrome browser, Google.com.

## GOOGLE.COM

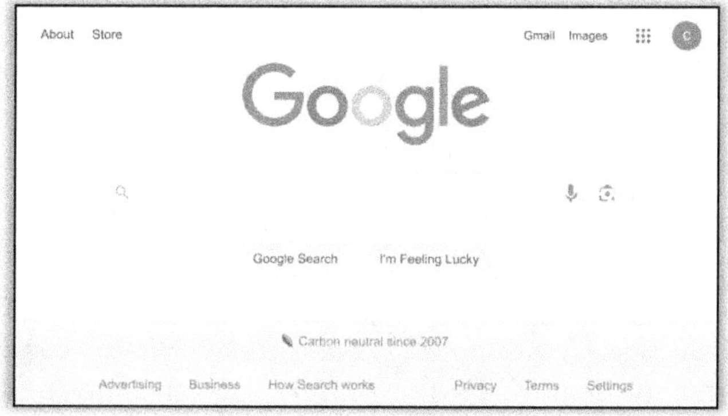


By some accounts, Google.com processes more than 8.5 billion searches per day. This is an enormous amount of data, putting Google.com as the most visited website in the world in February 2023, attracting nearly 97.0B monthly visits. With second-placed Youtube.com receiving 80.5B visits in the same period. That's more than 177B visits per month combined. That is a ton of traffic, and we access all this though Google.com. There is no limit to what you can search for on Google, and the Google search bar is a default widget on Android phones and is a downloadable widget on Apple products. Of course, all this power comes at a cost of $0 to us as the end users.

## GOOGLE ASSISTANT

Google Assistant is another way we access Google Search and Google.com. On most mobile devices, people tend to hit the microphone and say what they need or ask their question. This is done through the Google Assistant.

Similar to Alexa, the Google Assistant can initiate many tasks such as turning on lights, playing music or telling us the weather, but it is also used when you ask Google a question or start your search via the search bar microphone. When using your phone to get directions before a trip, you would start by asking your phone for directions using Google Assistant, which then would use Google Search to locate the destination and provide detailed information on multiple routes you could take, and then it could launch Google Maps to route you on your journey.

## GOOGLE MAPS

Google Maps is an incredible product that allows us to effortlessly navigate our surroundings. From travel time to alternate routes, Maps shows up to provide us with the data to make great choices when we are heading out the door. Remember the days of Tom Tom and paying for your gps service?

When you first launch Google Maps from your computer, you are presented with an overview map of where you presently are and some recommendations of things in the area.

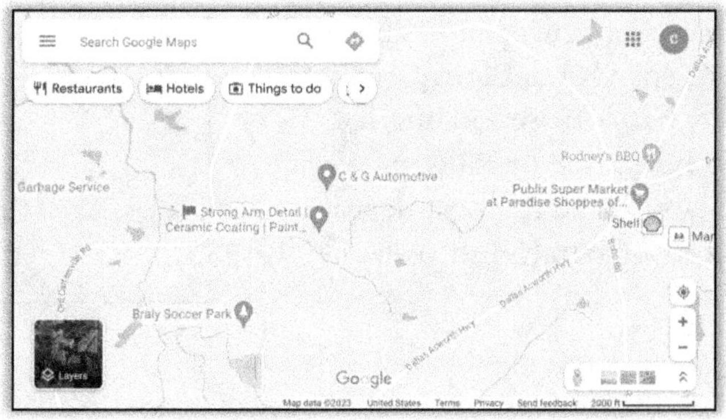


You can choose to search for your location in the top search bar or use one of the suggested categories to get recommendations for those types of businesses.

Maps is not only a complete navigation system, but it makes checking out your surroundings simple. You can zoom in on where your (location) pin is on the map and change the view to a satellite view. This allows you to familiarize yourself with the local landmarks and helps you to be comfortable in a new town or city. It is nice to see a familiar building or sign, so you know you are heading in the right direction.

When you are on your phone, at the bottom of the screen, you have options to Explore, Go, Contribute Updates or Business.

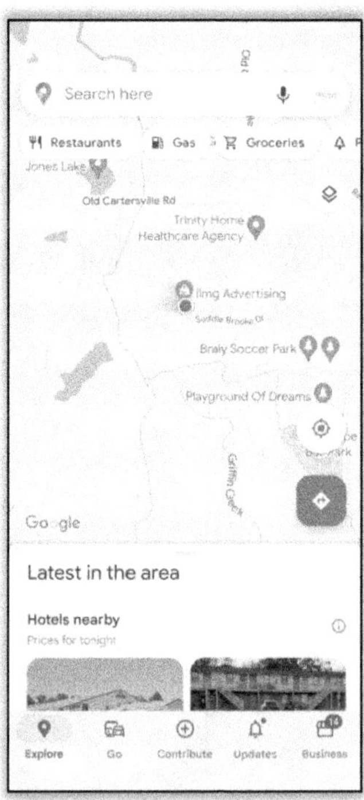


Remember this is where you come to work on your Google Business Profile from your mobile device.

As you see Google is helping us in the background of our lives every day. From the power of GPS to the ability to search your local area for exactly what you are looking for, Google has a free product or service to help you make the best decision for you, your family and your business.

1. Knowing Google Assistant is available. Do you have the search bar with the microphone on your phone's home screen?

______________________________________________

______________________________________________

2. Ask the Google Assistant some questions. List 3 things you asked and the reply you got.

______________________________________________

______________________________________________

______________________________________________

______________________________________________

______________________________________________

______________________________________________

______________________________________________

3. What is your home page on your computer browser?

______________________________________________

______________________________________________

4. Did you know you can use the Google assistant on your laptop or desktop computer?

______________________________________________

______________________________________________

5. List the last 3 places you used your navigation app to get directions to. Did you use Google Maps?

________________________________________

________________________________________

________________________________________

________________________________________

________________________________________

________________________________________

Remember, for updated information about this chapter, scan the QR code and watch the video for this chapter!!

Scan here to watch this chapters updated video

George's story continued

Over the years, George came to me to add a photo or change some wording on his website. He posted on Google Business consistently. He continued to build relationships with every client. He asked for reviews, every time! And he built his business. People read his reviews and would call him just to see if he really was that nice of a guy.

And he innovated. He found other complementary businesses that needed junk removal. Auto body shops, realtors, landscapers, anyone who might need to have something hauled off became his friend and an affiliated business. He would promote them to his clients, and they would use his services as well as refer his company to their clients, friends and business associates.

And his business grew. He hired other guys; he would subcontract with other junk companies and keep their workers busy when he was overwhelmed. And he gave it back. He paid off the child support and never fell behind again. He paid off the attorneys and never owed them again. He paid cash for his next truck.

He was living the American Dream!

# A FINAL NOTE:

# PUTTING IT ALL TOGETHER

"You don't have to see the whole staircase, just take the first step."

— Martin Luther King, Jr.

Now what? How do you take the changes you have made to your Google Business Profile, and the information about all the other free stuff you have learned from me in this book and leverage it to get the maximum amount of value out of it all? Are you just going to forget what you learned and hope for the best, or are you going to act?

If you did the 7-day process "Setting up your Google Business Profile" from chapter 4, you are already ahead of most businesses, but following up with all the other free products and comparing them to what you are currently using and how much it is costing you is a necessary next step to maximize the value of the Free S#!t Google has for you. Do your homework! This is money you are spending that you don't have to! I encourage you to investigate one of the free products I discussed in this book each week for the next two months. Just start at the beginning and work your way through the list of products. See if any are a good fit for your business and implement the ones that make sense.

I challenge you to act! Knowledge is not power. "Applied" knowledge is power. You can read all the diet books in the world but if you don't apply the wisdom in those diet books to your health and exercise regimen, you will never lose weight. If you never log into your Google Business Profile and post an update or some pictures, you will never get the attention of the Google algorithm and get your Business Profile elevated in the search results.

The choice is yours. How do you choose to interact with Google and promote your business? Will you give money to marketing companies who don't really know your business and hope they can let the rest of the world know how awesome you are, or will you invest some time making sure the most used search platform in the world knows what you do and loves your business profile?

This is a VERY important question, so choose wisely. There is no doubt that Google is the most used search platform in the world. The only question now is, will your business show up when people are looking for you?

In the ten exercise lines below, list out ten actions that you commit to taking within the next 90 days because of reading and applying the wisdom offered in this book. I did number 3 for you.

1. ________________________________________
2. ________________________________________
3. Posting on Google Business Profile 3 times a week.
4. ________________________________________
5. ________________________________________
6. ________________________________________
7. ________________________________________
8. ________________________________________
9. ________________________________________
10. ________________________________________

In this book you learned what a Google Business Profile is. You learned why you need a Google Business Profile. Also, what a Google Business Profile can do for your company. Learning the specifics about setting up your Google Business Profile and how Google views your company will allow you to get more algorithm love. Now you know all about SEO and when it is time to use Google Ads. Following the specific steps you have learned in this book will allow you to stay on top of the Google Search results. Act and stay motivated!

If you take the time to implement the step-by-step actions in this book, if you really put some thought into how your customer is looking for you, if you dig in and grasp how Google sees your business, then you will have a much higher chance for success with getting found online. Then you will start to see some of the savings in your operational budget, and that will help you get to six figures in your business sooner than later.

This is the real deal folks. This is not social media, and it is not a game. It can be fun, but it is serious business, and your business will benefit from you learning these principles and implementing them regularly and consistently. You deserve it, your business deserves it, and most of all the rest of the world is waiting for your great idea to make their lives better! Can I get an Amen?

So now that you have read my book, I encourage you to contact me and tell me what you liked about it. What really clicked with you or what made you scratch your head and wonder what I was trying to say? You will help me make this book better for the next printing and I can make sure I am bringing my best product to the world that is waiting for it!

But most importantly, tell me about you. If you want to share any challenges you are experiencing, any obstacles or adversities that are holding you back, I would be honored to try to help. In fact, I would like to offer you a complimentary, no obligation 30-minute consultation by phone, Zoom or in person (if geographically possible) to see how I can help and assist you.
My email address is Robin@RobinCStern.com, and my cell phone number is 678-685-9759, so email me, or better yet text me with your name and what time zone you are in, and we will schedule your complimentary consultation.

I wish you good luck! I wish you success and prosperity! I want you to take your amazing ideas to the world and get the viral algorithm activity that you have dreamed of, to elevate your business and exceed your wildest dreams for your future. Go out there and change the world, this is my challenge to you!

Your friend,

Robin

George's story continued

After 15 years of hard work and treating everyone as a friend, George was able to finally fulfill what he felt was his lifelong mission: To serve the homeless and overlooked populations in his local area. He broke his company up into multiple segments and sold each off to the person he felt would best continue his legacy.

He sold the company name, website and Google page to one of his dear friends from church, a fellow member of the "repurposed product" industry.

He sold the auto body scrap business to a very intelligent and hard-working young man for a percentage of future profits. That young man took that idea to levels of business and profitability that George never dreamed of, and he still faithfully pays George the percentage that was agreed to at the beginning (even though now the $$ amount has greatly increased due to the young man expanding the business)

He did buy himself a few nice things and helped his momma who always believed in him and supported him. And then he devoted his skills and resources to ministering to the homeless in his community. He takes necessary items to them regularly. He gives money when needed for medical issues or family needs. And he tells them about Jesus – the only thing George wants to do with the time he has left on this planet. To help others find the joy and freedom he found when he gave his life to his Savior.

I hope this story of George inspires you to go after what you feel you are called to do. To elevate your life to a level that will bring you joy and a sense of purpose and fulfillment.

I pray that you will use the things I shared with you to save the money that will help you take your business to the next level. In the words of Larry Z, my first business coach, "Your Business 2.0".

# EPILOGUE:
## KEEPING THE COOKIE FRESH

At the beginning of this book, we discussed your business and its relationship to Google when you are working on your Google Business Profile. This "freshest cookie" idea is an unbelievably valuable mindset to continue and carry forward in your dealings with Google and your business.

If you have ever come into the office breakroom and chomped into a 5-day-old chocolate chip cookie, you will start to see my point. You have to sneak into the breakroom and put fresh cookies there constantly. Dump the old coffee and make a fresh pot. Do you prefer the smell of 3-hour-old coffee on the burner and dried-up cookies or a fresh brewed cup and still warm baked goods?

What inspires Google's algorithm to want to put you in front of a customer? Why should they take a chance recommending you to the person searching? You can't get lazy. You can't brush it off. You have to take the initiative and step up to the plate every day and make the conscious decision to post an update, upload some pictures, ask for and answer a review. Promote, promote, promote.

This is your success or failure, your golden goose or rotten egg. Google is the platform where we all look. Will you show up? You cannot make a difference if you can't be found. It is not about money, it is about serving your tribe and changing the world.

In the following pages I have placed some more potentially helpful updates and ideas. Use them, don't use them. Just go out there and change the world one way or another!

# BONUS 1: KEEPING THE COOKIE FRESH WORKSHEET

Follow this worksheet regularly and you will be well on your way to making sure Google wants to put you at the top of your category! We have been working on this system for years with many of our clients and they are getting amazing results.

If you do nothing else please print this and post it on your wall, place it on your desk, enter these things in your calendar and keep it handy so you are reminded to do these things. This is the difference between stellar success and mediocrity. It's up to you.

Here is the link to this worksheet online. Use it!

https://yourgoogle.guru/wksht

# business profile maintenance guide

Maintaining your Google Business Profile can be overwhelming. Follow these steps to stay on track:

**yourgoogle.guru**

**Putting all these steps in your calendar to remind you will increase your success rate .**

**PHOTOS: ONCE A WEEK FOR 8 WEEKS**

- ☐ 20 random photo dump. All at once.
- ☐ Tags and quality are not important here
- ☐ Every 7 days to increase algorithm activity

**Make sure you know how to get to your Business Profile from mobile device**

**SEARCH: MY BUSINESS ON DESKTOP / USE MAPS APP ON MOBILE**

- ☐ Business link at bottom of app screen
- ☐ Reply to messages and reviews from profile
- ☐ Check performance and accuracy of profile

**Keep activity up with posting updates**

**UPDATES 3 TIMES PER WEEK MINIMUM**

- ☐ In Maps app: See business profile link
- ☐ Updates tab at top then select add update
- ☐ This is to generate activity / let Google know you are open for business

**Get Reviews, Get Reviews, Get Reviews!! Did I say you should get reviews?**

**REVIEW LINK ON WEBSITE**

- ☐ Update website with Review us button
- ☐ POST ABOUT NEW PROFILE ON ALL SOCIAL
- ☐ Request reviews from all past/current customers and all friends / family

·2023·
Grow with Google
HIGH IMPACT PARTNER

# BONUS 2:
# 10 GREAT IDEAS FOR YOUR NEXT POST

- Showcase Customer Testimonials: Share positive reviews and testimonials from satisfied customers on all your social pages to build trust and credibility.

- Behind-the-Scenes: Provide a glimpse into your company's daily operations or share stories about your team members to humanize your brand.

- Promote Exclusive Offers: Offer special discounts, promotions, or limited time deals to encourage customers to engage with your business.

- Share Success Stories: Highlight success stories of customers who have benefited from your products or services. Include photos or videos for added impact.

- Highlight New Products or Services: Introduce and showcase any new products or services you are offering, explaining their features and benefits.

- Celebrate Milestones: Share significant milestones or anniversaries your business has achieved, expressing gratitude for customers' support.

- Expert Tips or Advice: Share valuable tips, tricks, or industry-related advice to position your business as a knowledgeable resource.

- Community Involvement: Highlight your business's involvement in local charities, events, or initiatives, demonstrating your commitment to the community.

- Employee Spotlights: Spotlight individual employees, sharing their expertise, achievements, or contributions to your company's success.

- Interactive Polls or Surveys: Engage your audience by posting interactive polls or surveys to gather feedback or preferences on various topics related to your business.

Remember to tailor these ideas to suit your specific business and audience, and feel free to add your own creative touch to make them more compelling.

# BONUS 3: MORE FREE S#!T: BUSINESS TRAINING FROM GOOGLE

Head over to https://smallbusiness.withgoogle.com and you will find an incredible amount of content and information to benefit your business.

- Get the essentials: this will walk you through each step of getting online from your domain name, email address, website and of course requesting your free Google Business Profile
- You can book a session with a Google expert who can walk you through how to make the most of Google products to achieve your goals.
- Get a guide tailored to your business type and discover more ways to stand out and do business online, reach more customers, and work smarter with tools from Google.
- Step-by-step video tips and lessons to help you to succeed online.
- Building your online video presence with YouTube. This is a vital part of any business and Google owns YouTube so of course they have a great amount of info on how to get your business on YouTube and how to use that to reach your customers and audiences.

Here is an example from Google's Business Primer website smallbusiness.withgoogle.com:

## BEST PRACTICES
## FOR YOUR WEBSITE:

SEO is about using the right words and serving relevant information to your potential customers on your website. Here are some tips and best practices to get started.

- Provide useful product or service information.
- Use short and meaningful page titles and headings.
- Include details about your business such as your contact information, address and hours.
- Check image and video tags to make sure they are descriptive.
- Use descriptions that your potential customers can relate to.

# BONUS 4:
# 6 GREAT IDEAS TO JUMP START GETTING MORE REVIEWS

Sure a happy customer is the best advertisement and getting that customer to leave a review is a sure-fire way to maximize the value of all the effort you put into making that customer a 5-star reviewer, but there are other ways to get reviews. Here are a few of our favorite ideas from over the years:

1. Leave yourself a review. Seems simple but that is one more review. Google is ok with an owner telling the community they are ready to serve their neighbors and friends.

2. An employee telling potential customers how great it is to work at your establishment. They can be very clear in their review that they work there. This is great info for Google to share with their customers.

3. Ask past customers. If you have people you have worked with in the past, you can ask them to give you a review. Even if you are in a different position or business, if they review your character and traits from when they worked with you or your company, Google sees that as valid information.

4. Character references. If you have some folks that know you personally. If you have served on a nonprofit board or in some public capacity and folks would be willing to put it in writing, send them a link!

5. Make sure you put your review link on your website as a button requesting reviews. This will help when clients don't respond immediately but remember later. They can head to your website and find the button there. This helps with page traffic to your website, as we discussed in a previous chapter of this book.

6. Ask. Put up a sign in your shop if you have a brick and mortar. Put it on your business card. Put a QR code linking to your website on your vehicle and ask folks to scan it. Whatever you have to do to get the word out or remind yourself to ask, do that!

# BONUS 5: HOW TO HASHTAG YOUR A$$ OFF

Creating a unique connection to your brand is critical in showing up when your potential clients and raving fans are looking for you. Hashtags are a great way to do this.

Most of us know that we can use hashtags to connect to a larger group. If I post on social media and put the hashtag #google, you can be sure I will get a lot of folks seeing my content if they are looking for posts related to Google. This is also a great way to get Google to notice you.

Another cool way to use hashtags is to get your own and use it to aggregate all your content and information. This is how you would go about that.

Let's say your company name is Bobby's Lawn and Landscape. Now if you Google that with a hashtag in front of it you would get one search result for a Facebook page for Bobby's Lawn and Landscape. Not good. But if you add your last name initial, you would Google #bobbybslawnandlandscape and you would see this as your search result:

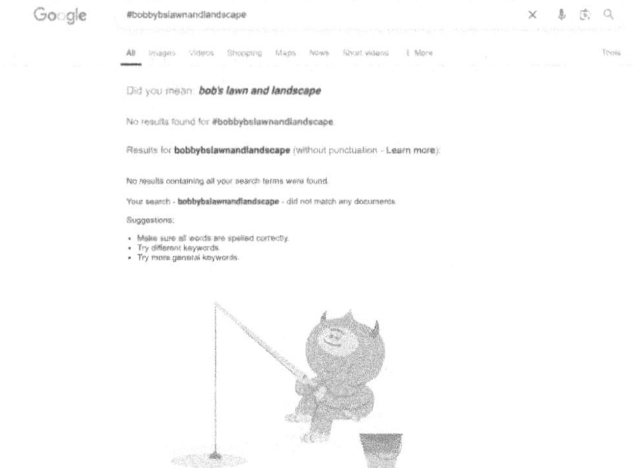


This is the hashtag you would want to use. Now don't get me wrong, this is a long hashtag and typing it on every social media post, comment and share could get overwhelming, so we suggest using something shorter, but for the purpose of our example, I wanted you to see my little ice fishing troll. He is the one that lets you know Google has no good results for that hashtag, and that is exactly what you want.

Nobody owns a hashtag, but if you find a unique one that matches your brand, that will allow you to aggregate all your content under that one hashtag and then when people search your hashtag, you stick out like a rockstar with only your search results showing for that hashtag. Not a bad way to own some real estate on the internet for the amazing price of…. NOTHING!!

# ABOUT THE AUTHOR

Robin C. Stern is a visionary author, professional keynote speaker, educator, and technology guru with an insatiable curiosity for exploring the ever-evolving world of digital tools. As a leading expert in harnessing the power of technology, Robin's passion lies in empowering readers to make the most of freely available resources, and his latest masterpiece, "Using Google's Free S#!t," is a testament to his dedication to democratizing the digital realm.

A trailblazer in the tech landscape, his journey into the digital world began with an unquenchable thirst for knowledge. With an exceptional gift for communication, he quickly realized the potential of technology as a force for creativity and resourcefulness. Through years of hands-on experience and exploration, he has honed his skills as an outstanding communicator and educator, making complex concepts accessible and exciting to learners of all levels.

In addition to his expertise in the digital realm, Robin owns iluvmygeek, a brick-and-mortar computer store in Dallas, GA, which allows him to connect with the local community and provide personalized technology solutions to individuals and businesses alike. This experience further enriches his understanding of the diverse applications of technology in everyday life.

His journey is one that embodies a true sense of service and dedication. Prior to his ventures in the technology industry, he served in the US Marines, instilling in him the values of discipline, integrity, and commitment, which continue to shape his work ethic and approach to life.

"Using Google's Free S#!t" is a groundbreaking work that showcases Robin's visionary perspective on digital resources. In this book, he reveals how Google's suite of free tools can be harnessed to transform personal, educational, and professional endeavors. From unlocking the secrets of Google Calendar to seamlessly collaborating through Google Docs and harnessing the power of data analysis with Google Sheets, Robin leaves no stone unturned in helping readers unlock the true potential of these invaluable resources.

Beyond his writing and teaching endeavors, Robin C. Stern remains a driving force in the technology community. His dynamic presence as a sought-after speaker at tech conferences and seminars sparks inspiration and fuels a passion for embracing digital solutions with confidence.

While his mastery lies in the digital domain, he also cherishes the beauty of the natural world. Often, he can be found seeking solace on hiking trails, where his creative ideas find harmony with the serenity of nature.

In conclusion, Robin C. Stern's extraordinary journey as a technology enthusiast, exceptional communicator, entrepreneur, and military veteran makes him a true pioneer in bridging the gap between complex digital tools and everyday users. His dedication to empowering readers with the knowledge to navigate the digital landscape sets him apart as a thought leader and a friend, guiding individuals and organizations towards embracing the boundless opportunities that technology offers. Through his captivating words and insights, Robin invites readers to embark on a transformative journey of digital discovery, leading them to a future where technology becomes a powerful ally in realizing their dreams.

For more information, visit: https://robincstern.com

# ABOUT ROBIN C. STERN COACHING

If you want to stand out from your competition, leverage the free products and services from Google, and get free publicity, then perfecting your Google Business Profile and making sure Google loves you and wants to put you higher in their search results is top priority.

Don't feel overwhelmed by the thought of setting up and maintaining your Google Business Profile. Let leading Google Business Profile Partner Robin C. Stern show you the way.

If you have struggled with:

- Understanding how you interact with Google products.
- Making sure Google loves your Profile page.
- Driving more traffic to your website
- Getting reviews
- Increasing revenues
- Increasing online engagement and activity
- Selling your stuff online – and getting sales!

You have come to the right place, and NOW is the time to act.

Robin C. Stern is offering you a complimentary, no obligation 30-minute consultation by phone, Google Meet, or in person if geographically possible. Act now, your future success is waiting!

His email address is Robin@RobinCStern.com and his cell phone number is 678-685-9759. Please email, or better yet text, with your name and what time zone you are in, and we will schedule your complimentary consultation.

www.RobinCStern.com
Robin@RobinCStern.com
Mobile: 678-685-9759

# ABOUT ROBIN C. STERN CONSULTING

Every company, regardless of its size or industry, can benefit from the expertise of Robin C. Stern, your highly skilled Google consultant. With his deep knowledge of Google's free products and services, Robin can help any company save money while increasing its online visibility.

By leveraging tools like Google Business Profile, Google Analytics, and Google Search Console, Robin empowers businesses to optimize their digital strategy, attract more customers, and drive results. With his cost-effective approach and proven track record, Robin C. Stern is the ideal partner to help any company thrive in the competitive online landscape.

Robin's value lies not only in his expertise but also in his ability to tailor his services to meet the unique needs of any company. He takes the time to understand the specific goals and challenges of each organization, providing personalized guidance and support.

Robin's comprehensive training and actionable steps ensure that companies can independently leverage Google's free products and services for long-term success. Whether a startup or an established enterprise, any company can benefit from the invaluable insights and cost-saving strategies offered by Robin C. Stern, YOUR trusted Google Business consultant.

Robin C. Stern is offering you a complimentary, no obligation 30-minute consultation by phone, Google Meet, or in person if geographically possible. Act now, your future success is waiting!

His email address is Robin@RobinCStern.com and his cell phone number is 678-685-9759. Please email, or better yet text, with your name and what time zone you are in, and we will schedule your complimentary consultation.

www.RobinCStern.com
Robin@RobinCStern.com
Mobile: 678-685-9759

# ABOUT ILUVMYGEEK

In 2018 with 3 employees and outgrowing the basement, I decided (with the assistance and encouragement of my employees) to take the leap and move our successful computer business into a brick-and-mortar storefront / service center. This was no small task, but as the stars aligned, we were able to put the whole thing together in a cool little building we had some previous history with.

About 3 miles from our original location (my house) in Dallas, Ga, we found a gentleman who was operating a drive through emissions inspection station. (Yes, those still exist in this country). He had a nice sized building that had a lot of vehicles passing daily so I approached him and asked if he would be interested in sub leasing the space that he wasn't using. He had attached a small aluminum building on one end of the structure for the cars to drive through to be inspected, but he was not using much of the inside of the building.

There was some discussion of how we could divide up the space to accommodate both businesses and we struck a deal! We built a wall to divide the space and put up the signs. We changed our Google Business Profile information, and we went to work! I can honestly say that I am living proof that Google Business works. We have built that business to more than 600 + Google Reviews, and we have had a minimum of

25% growth year over year since we moved into our retail space.

This is important. We rarely do any other forms of advertising. We sometimes boost a FB post if it gets a lot of engagement, and we have been known to pay to be a vendor at a local event, but as far as other forms of advertisement, we rely almost exclusively on our Google Business Profile listing and Google reviews. We just had the best month in the existence of the company (August 2023) and we are excited for the opportunity to help more customers and local families to be able to navigate the computer world with less stress and confusion. Thank you, Google, for allowing us to get the message out and serve our community.

I mentioned earlier that we had some previous history with our building. When we moved to this area, 12 years before opening that store it was a Taekwondo dojo. All my boys practiced this marital art in that building, including my oldest son Tyler who is now running that business and works in that very same building every day! Becoming a part of your community is a large piece of any successful business, so if opportunity presents itself, go for it and take every chance to grow your influence and local reputation!

# ABOUT "ASK YOUR GOOGLE GURU" INNER CIRCLE

Keeping up with all the changes as well as keeping motivated to stay the course is the focus of the Ask Your Google Guru Inner Circle. We utilize the Book of Face (and appreciate the platform) so we can stay in touch and keep you on track.

Are you ready to take your Google knowledge and expertise to the next level? Look no further than the "Ask Your Google Guru Inner Circle" Facebook group - a tight-knit community of passionate individuals who are committed to unraveling the depths of Google's vast universe. By joining this exclusive group, you join a network of like-minded individuals who share your insatiable thirst for digital knowledge and innovation.

Inside the "Ask Your Google Guru Inner Circle," you'll find a treasure trove of insider tips, advanced techniques, and cutting-edge strategies to master Google's arsenal of tools and services. Engage in discussions with industry experts, thought leaders, and fellow enthusiasts who are at the forefront of the digital revolution. This is the perfect space to exchange ideas, seek personalized advice, and stay ahead of the curve in an ever-evolving technological landscape.

Don't miss your chance to surround yourself with the best and brightest minds in the Google community—join the "Ask Your Google Guru Inner Circle" today and unlock unparalleled opportunities for growth, networking, and innovation.

For more information on Ask Your Google Guru Inner Circle, visit: https://robincstern.com/services

# BOOK ROBIN C. STERN TO SPEAK AT YOUR NEXT EVENT

When it comes to choosing a professional speaker for your next event, you will find no one more respected or successful - no one who will leave your audience or colleagues with a more renewed passion for promotion - than Robin C. Stern, one of the most gifted speakers of our generation.

Since 2016, Robin has delivered many presentations locally and virtually all over the world in international summits. Whether your audience is 10 or 10,000, he can deliver a customized message to inspire your meeting or conference.

Robin understands your audience does not want to be "taught" anything, but is rather interested in hearing stories of inspiration, achievement, and real-life people growing their businesses organically and with proven principles that are simple and reproducible.

As a result, Robin's speaking philosophy is to provide humor, entertainment, and inspiration to your audience with passion and stories proven to help people achieve extraordinary results by understanding and applying the concepts of online marketing.

If you are looking for a memorable speaker who will leave your audience wanting more, book Robin C. Stern today!

To see a highlight video and find out whether he is available for your next meeting, visit his site at the address below. Then contact him by phone or email to schedule a complimentary pre-speech phone interview:

www.RobinCStern.com
Robin@RobinCStern.com
Mobile: 678-685-9759

www.ingramcontent.com/pod-product-compliance
Lightning Source LLC
LaVergne TN
LVHW010609100826
845148LV00014B/2904